# IN THE LIGHT OF THE SUN

Trees, Wood, Photosynthesis and Climate Change

# John Halkett

Foreword by Dr René Castro Salazar

**Connor Court Publishing Pty Ltd**

Copyright © John Halkett 2018

ALL RIGHTS RESERVED. This book contains material protected under International and Federal Copyright Laws and Treaties. Any unauthorised reprint or use of this material is prohibited. No part of this book may be reproduced or transmitted in any form or by any means, electronic or mechanical, including photocopying, recording, or by any information storage and retrieval system without express written permission from the publisher.

PO Box 7257
Redland Bay, Qld 4165

sales@connorcourt.com

www.connorcourtpublishing.com.au

ISBN: 9781925501810

Interior design and illustrations by Russell Jeffery

Cover design by Ian James

Printed in Australia

For Maxeine, the trees and the survival of humanity on planet Earth.

# Other books by John Halkett

*The World of the Kauri* (with E V Sale)
Reed Methuen, Auckland, 1986.

*The Native Forests of New Zealand*
GP Books, Wellington, 1991.

*Tree People* (with Peter Berg and Brian Mackrell)
GP Books, Wellington, 1991.

*Trees that call Australia home*
Potts Point Publishing, Sydney, NSW, 2009.

*Jungle Jive: Sustaining the forests of Southeast Asia,*
Connor Court Publishing, Brisbane, 2016.

# Contents

| | |
|---|---:|
| Foreword René Castro-Salazar | vii |
| Getting started | xi |
| Chapter 1<br>Setting the scene: The need for courage in the face of pessimism | 1 |
| Chapter 2<br>Weather – but not as we know it: And you thought a few thousand Syrian refugees created a drama | 16 |
| Chapter 3<br>People by the billion: Population explosion put pressure on the planet | 28 |
| Chapter 4<br>Feeling the heat: The need to tackle climate change has become urgent | 37 |
| Chapter 5<br>Living in a greenhouse: Things are hotting up fast and natural causes cannot account for such rapid warning | 46 |
| Chapter 6<br>Oh no – not a chemistry lesson: Calculating a safe carbon budget for humanity. | 53 |
| Chapter 7<br>Nature's magic show: How in a process called photosynthesis leaves turn carbon dioxide from the sky into giant trees | 64 |
| Chapter 8<br>So what about forests then? A key ingredient in balancing carbon accounts | 76 |

**Chapter 9**
Selling the carbon to save the trees: Emissions trading – fighting climate change and protecting forests — **88**

**Chapter 10**
Let's work together; Global agreements setting the pace — **95**

**Chapter 11**
Getting the good wood message: Employing the power of photosynthesis — **101**

**Chapter 12**
Feeling better now? Measurable health and wellbeing benefits if you have wood in your life — **113**

**Chapter 13**
Is the sky the limit? Inwards and upwards, drivers behind city building innovation. — **124**

**Chapter 14**
An international phenomenon: Tall buildings with the resilience of trees — **145**

**Chapter 15**
More wood wonders: Biology and renewability keys to continuing benefits — **155**

**Chapter 16**
Burning wood: Exploring the contradiction that burning trees fights climate change — **166**

**Chapter 17**
Action stations: Harnessing the power of photosynthesis-derived products — **179**

Acknowledgements — **199**

Picture credits — **201**

Index — **203**

# Foreword

René Castro Salazar[1]

I am pleased to have been invited to contribute a foreword to this book, the subject of which is important to my role at FAO[2] and one I am passionate about. Absolutely no doubt, climate change is the biggest challenge facing the continuing survival of humanity here on Earth. It is a topic that consumes a substantial amount of my professional life and personal thoughts. Also in past roles, both as a minister in the Costa Rican Government and as a Fellow at Harvard University, I put energy into policies, practices and arguments related to climate change abatement.

If I can say so here, I am very proud of the leading role Costa Rica has, and continues to play in leading climate change policies and practices. In 2007 Costa Rica set itself an ambitious climate change mitigation target – to become carbon neutral by the year 2021. This means that the net balance of greenhouse gases released into the atmosphere by Costa Rica from 2021 onwards would be zero.

This does not mean that the country will have zero greenhouse gas emissions – rather it means that emissions will be reduced to the point where offsets are sufficient to cancel out all remaining emissions so as to achieve a net greenhouse gas contribution of zero. Costa Rica's adoption of this target reflects the view that carbon neutrality could be achieved without compromising the country's national development goals.

---

1   Dr René Castro Salazar was appointed Assistant Director-General for FAO, Forestry Department, and in December 2016 he was appointed Assistant Director-General for the new Climate, Biodiversity, Land and Water Department. Dr Castro Salazar held ministerial positions in Costa Rica between 1994 and 2014 as Minister of National Resources, Minister of Foreign Affairs and Minister of Environment and Energy between 1994 and 2014. He has also been a Fellow at Harvard University.
2   Food and Agricultural Organization of the United Nations

René Castro-Salazar: Firmly and forcefully of the view that forests are at the heart of a transition to low-carbon economies.

In April 2014 I was able to report that 81 per cent of Costa Rica's progress toward carbon neutrality to that time had been achieved through the offsetting effects of natural forest regeneration. Under its successful forestry policy Costa Rica has become a pioneer tropical country to invert its rate of deforestation, increasing its forest cover from 21 per cent in 1987 to 52 per cent in 2013. This has been internationally acknowledged as a stunning achievement, the result of a sustained social and economic effort, including a fuel tax that has contributed 80 per cent of the $500 million invested in reforestation between 1996 and 2013.

It is now crystal clear to me that the agriculture, forestry and land-use sectors have the greatest potential for carbon sequestration and offer abatement opportunities in the short term. They are cheaper than those in the energy or transport sectors. In two or three decades technology development may change this competitive advantage of the forest based activities and projects, but not today.

John has long been an enthusiastic advocate of forest conservation

which forests increase removals of carbon from the atmosphere, while sustainable forest management and wood products contribute to enhanced livelihoods and a lower carbon 'footprint'.

Forests do sterling work in 'locking-up' carbon dioxide. The relative speed and cost-effectiveness with which forests make their presence – or absence – felt is a key reason why they figure prominently in the plans countries are crafting to meet commitments made in the Paris Agreement on climate change.

Where forests are sustainably managed and relatively abundant, woody biomass – often made from recycled or waste products – can serve as a large-scale energy source. Promoting wood as a renewable energy source may seem counter-intuitive, but almost two billion cubic metres of wood – more than half the world's wood output – is already used for that purpose, highlighting the potential gains from more sustainable forest management. So clearly forests are at the heart of the transition to low-carbon economies, including the wider use of wood products to displace more fossil fuel intense products.

More directly, when wood is transformed into furniture, floors, doors or beams it continues to store carbon for decades. Today there is growing evidence that wood-based products are highly competitive with alternative construction materials. The carbon balance of a timber-frame building is only half that of a concrete frame structure.

Enough from me here – I commend John's book to you as a well-considered, informative and easy reading account of the extremely valuable link between trees, wood, photosynthesis and combating climate change.

and sensitive management. His last book *Jungle Jive: Sustaining the forests of Southeast Asia* sits comfortably with the outcomes FAO would like to achieve in relation to tropical forests.

Whilst across temperate regions of the world forest and tree plantation cover is expanding, in *Jungle Jive* John points to the continuing dark stain of deforestation and forest degradation in tropical forest regions. This serious challenge needs urgent attention, and continually exercises our minds at FAO – both for the forests themselves and their inhabitants – plus as a tool to assist climate change abatement.

John advocates that addressing the seemingly irresponsible and irrational tropical forest destruction seen across South East Asia needs an economic approach that makes forests too valuable to destroy. This argument is founded on the increasing prospects for sustainable, legally verified wood production, value-added manufacturing, plus climate change abatement carbon credit creation and trading.

John argues that such a prescription will help to create a set of circumstances where tropical forests are seen as economic assets, not liabilities, and where governments, corporations and local communities have a vested interest in keeping trees standing.

Turning more directly to this book, we know that climate change is the key challenge of present and future generations. The impact of increased global temperature will affect all regions and countries, but will hit hardest those already living in poverty and with food insecurity. It will badly affect tropical countries that are more biologically diverse and less adaptive to quick changes.

I am firmly and forcefully of the view that forests are at the heart of a transition to low-carbon economies. Forests and wood products have a key role in climate change mitigation and adaptation through the potential for wider use of wood products to displace more fossil fuel intense products. Indeed, a virtuous cycle can be enacted in

# Getting started

I don't know about you, but I'm a tree guy, both work-wise and privately. I have always enjoyed the company of trees and prefer to visit a forest or tree plantation rather than say an art gallery or museum.

Also I am proudly a forester. Although traditionally more associated with game-keeping, in the seventeenth century foresters began to measure trees, assess growth rates, and calculate, in accordance with the evolving concept of sustained yield management, prescribe annual allowable log harvesting levels. So I have a distinguished history and forest conservation pedigree to honour and perpetuate.

As stated in Chapter 1, the science behind climate change is

Author John Halkett: A distinguished history and forest conservation pedigree to honour and perpetuate.

comprehensively settled, so I am not about to debate whether or not the climate is changing. As far as I am concerned, the scientific argument is robust and the evidence overwhelmingly compelling. There is unequivocally no question that the level of major atmospheric pollutant, carbon dioxide is rising. Science concludes that if this trend continues the implications for the climate, and for all life forms will be dire.

Even if the case wasn't so compelling, you would think that on a precautionary basis, given that we have nowhere else in the galaxy to live, we would err on the conservative side of the argument. Looking after our home here on planet Earth, and ensuring that it continues to remain habitable for all residents seems to be the obvious and only direction for us to adopt.

Back to trees and forests for a moment, I consider the conservation and management of natural forest systems and tree plantations as being central to a sustainable and low carbon future for humanity. So I don't mind being called a 'greenie-with-a-chainsaw'.

Yes – cutting down and utilising the right tree in the right place is an expression of the value and respect humanity places on the qualities and attributes of trees. Our survival may well depend on a better understanding and use of trees in ways that only now are being contemplated – as a means of mitigating climate change, as a source of energy, renewable fuels, chemicals and other products yet to be discovered. As the twenty-first century marches on this will become vitally important, particularly so as trees and forests come under constant attack as human populations continue to grow like topsy. As climate change impacts bite, sea levels rise, and volatile weather events threaten to overrun humanity's existence, trees, forests and wood products are and will continue to be even more important.

Cutting down and utilising the right tree in the right place is an expression of the value and respect humanity places on the qualities and attributes of trees.

In the chapters ahead you might sense that I am a bit fixated by the essence of life on Earth – the mysterious process of photosynthesis. This fascinating piece of chemistry is central to the growth of trees and other plants and to all life, and to the fight against climate change.

In the pages ahead I argue that we must look after the forests we have. I assert that we need to rehabilitate degraded forest, tackle deforestation, plant more trees and use more renewable, carbon neutral wood. You will read that I put the proposition that sensitive timber harvesting is good – carbon stored off site – plus more stored in new regenerating trees. So I believe the case for sustainable forest management and harvesting is a key aspect of the case I attempt to advance.

---

This book follows on somewhat from my last – yes another book with a focus on trees and forests – *Jungle Jive: Sustaining the Forests of Southeast Asia*. Put simply this book argues that we need to better recognise the incredible values of Southeast Asia's tropical forests, and figure out what we can do to ensure that these values are safeguarded so that the plants and animals calling these forests home are sustained for their own and for our sake.

The book also flags the implications from the emerging adverse impacts of climate change and the need to reduce – or offset – carbon dioxide emissions. It discusses the role tropical forest retention can play in assisting to mitigate adverse climate change affects. So to some degree, this book follows on from the proposition I have put in *Jungle Jive: Sustaining the forests of Southeast Asia* about the importance of perpetuating the jungles of Southeast Asia for a whole range of reasons, including climate change abatement.

Borneo orangutan: A need to better recognise the incredible values of tropical forests, and figure out what we can do to ensure that these values are safeguarded.

Getting started  xv

In the context of photosynthesis, there are practical, cost-effective measures available right now to confront climate change adversities.

---

In this book I back up my case by drawing on the reputation and research excellence of the Food and Agricultural Organisation (FAO) of the United Nations. This entity points out that forests are at the heart of the necessary transition to a 'low-carbon' economy. This not only because of the role forests play as a 'carbon sink', but also through the prospective wider use of wood products to displace more energy intense products. We will define these terms and discuss them and related issues in the pages ahead.

So you don't need to take it from me – FAO points out that forests do *herculean* (their word) work in locking carbon dioxide into leaves, branches and soils. The relative speed and cost-effectiveness with which forests make their presence – or absence – felt is one key reason why trees and forests figure prominently in the plans countries have crafted to meet commitments made in the 2015 Paris Agreement on climate change. Again we will detail this and other relevant global environmental agreements later.

---•---

I have been conscious not to make this book too hand-ringing and despondent, even though there is much to be worried about climate change-wise. Rather, I have suggested, at least in the context of photosynthesis, that there are practical, cost-effective measures available right now to confront climate change adversities.

It is probably not too much of an exaggeration to say libraries of books have already been written about the science of climate change. The subject is complex, so trying to make it more digestible has been one of the challenges in writing this book. Equally, the chemistry behind photosynthesis is somewhat mysterious and complicated. Again, I have attempted to simply the chemistry and language. I hope that I have, at least in part, moved in this direction. See what you think.

CHAPTER 1

# Setting the scene

The need for courage in the face of pessimism

You don't have to be a genius, a rocket scientist – or a climate change expert for that matter – to work out that all the pollutants from the millions and millions of tonnes of fossil fuel we have burnt since the Industrial Revolution[3] has gone up there into the atmosphere somewhere, and cannot be doing us or the planet much good. Not hard to figure out. What do you think?

Coal-fired power plant emitting smoke over cityscape. Millions of tonnes of pollutants from the burning of fossil fuel since the Industrial Revolution have gone up into the atmosphere.

---

3   The Industrial Revolution was considered to be the time of transition to mechanised manufacturing that occurred in the period from about 1760 to about 1840. This period included moving from hand production methods to machines, new chemical manufacturing and iron production processes, the increasing use of steam power and the development of mechanised tools. The transition also included the change from wood to coal and the start of oil-based fuels.

**2** By the light of the Sun

This book is certainly not going to debate whether or not the climate is changing. As far as I am concerned, the scientific argument is settled and the evidence compelling. However, we will recap why the climate is changing and the influences that have caused this change. However, we will concentrate most of our efforts on what we might do about this situation.

Sure – debate continues over precisely by how much global temperatures might rise as climate change-related atmospheric pollution continues. But there is now unequivocally no question that levels of the major atmospheric pollutant, carbon dioxide are rising, and that this trend is, and will continue to have dire implications for the climate, and for all life forms that call planet Earth home.

No longer a scientific curiosity, adverse climate change is now the overriding environmental issue facing our survival. It will continue

Polar bear navigating between ice floats in the Arctic Circle, Barentsoya, Svalbard, Norway. Rising atmospheric carbon dioxide levels will continue to have dire implications for the climate, and for all life forms.

Environmental activist and author David Suzuki speaking on the eve of the Paris climate summit, promoting a vision for a sustainable future.

to affect our society in economic, health, food production, water supply, international security and in other aspects of our daily lives.

In his book *The Legacy: An Elder's Vision for our Sustainably Future*[4], David Suzuki provides a vision for a sustainable future. In relation to humanity's climate change and other environment impacts on planet Earth he says[5]:

> *Only by confronting the enormity and unsustainability of our impact on the biosphere will we take the search for alternative ways to live as seriously as we must. As an elder, I am impelled by a sense of urgency that comes from the recognition that my generation has induced change and created problems that we bequeath to my children and grandchildren and all generations to come. That is not right, but I believe that it is not too late to take another path.*

---

4  *The Legacy: An Elder's Vision for our Sustainably Future*, David Suzuki, 2013 Allen & Unwin, Vancouver, BC, Canada.

5  *The Legacy: An Elder's Vision for our Sustainably Future*, Page 3.

Reinforcing David Suzuki's testimony, in a 2016 report the FAO[6] says:

> Human-induced climate change poses one of the greatest challenges of the twenty-first century. Current atmospheric concentrations of carbon dioxide ($CO_2$), methane ($CH_4$) and nitrous oxide ($N_2O$) are unprecedented in at least the last 800, 000 years. The daily average concentration of $CO_2$ in the atmosphere rose above 400 parts per million (ppm) for the first time on record in 2013, up from 280 ppm before the Industrial Revolution and 315 ppm[7] when continuous observations began at Mauna Loa in the United States in 1958. Greenhouse gas (GHG) emissions continue to increase. As the $CO_2$ concentration in the atmosphere rapidly approaches 450ppm, it has become clear that limiting the global temperature increase to a maximum of 2°C requires urgent and comprehensive actions from most important sectors in the major emitting countries, irrespective of their current economic status.

Most importantly this book will attempt to outline what we can do to slow, and eventually halt this human-induced climate change juggernaut. In the pages ahead we will focus in particular on the process of photosynthesis, and the central role trees and forests, plus wood, are capable of playing in the fight to arrest catastrophic climate change.

---

As we will see, not only do trees 'suck' carbon dioxide out of the atmosphere and wood products store it away for decades or longer, but trees also have the capacity to be at the centre of a sustainable future for humanity. Yes – trees will be an essential element beyond the end of the fossil fuel era. They will be a critical ingredient in searching for a zero net carbon emissions future necessary if the planet is going to remain habitable for my three grandkids, and if you are lucky enough to have any, yours as well.

Boiled right down, climate change is the direct result of air

---

6  *Forestry for a low-carbon future: Integrating forests and wood products in climate change strategies 2016* Food and Agriculture Organization of the United Nations (FAO), Rome.

7  ppm is parts per million.

Setting the scene 5

Trees 'suck' carbon dioxide out of the atmosphere and wood products store it. Trees have the capacity to be at the centre of a sustainable future for humanity.

pollution. We know the size of our atmosphere and the volume of carbon-based and other pollutants that we are pumping into it, so the climate change debate now concerns the impacts of these pollutants on the continuation of life on Earth. What is it we are going to do to halt a continuation of this pollution? And what alternative energy sources and zero pollution solutions can be developed for tomorrow's global community?

---

The turbulence in public and media discussions suggests it has proven challenging for us to consider climate change dispassionately because of its political, commercial and industrial impacts, together with its implications for our personal economic welfare. This means that, as we seek to address adverse climate change, the probable consequences for short-term life style choices become a significant concern. This has led to a proliferation of misleading science and scary tactics as special interest groups argue their case for essentially ignoring the science and doing nothing. This uncertainty has created opportunities for political mischief, policy inertia and public confusion.

Commentary about climate change denial is becoming more strident. In an opinion piece in the *New Zealand Herald* Jarrod Gilbert[8] suggests that climate change denial should be a criminal offence. He says:

> There is no greater crime being perpetuated on future generations than that committed by those who deny climate change. The scientific consensus is so overwhelming that to argue against it is to perpetuate a dangerous fraud. Denial has become a yardstick by which intelligence can be tested. The term climates sceptic is now interchangeable with the term mindless fool. ... There may be differing opinions on what policies to pursue, but those who deny that climate change exists ought be shouted down like the charlatans that they are. Or better yet, looked upon with pitiful contempt and completely ignored.

In more measured terms, English physicist Professor Brian Cox, whose books and television programmes explain complex scientific phenomena in accessible ways says ignoring the best evidence is:

> ...entirely wrong, and it's the road back to the cave. The way we got out of the caves and into modern civilisation is through the process of understanding and thinking. Those things were not done by gut instinct. Being an expert does not mean that you are someone with a vested interest in something; it means you spend your life studying something. You're not necessarily right – but you're more likely to be right than someone who's not spent their life studying it.

Commenting on the severe damaging storm that savagely impacted on the East Coast of Australia in 2016 Elizabeth Farrelly[9] said[10]:

---

8  Dr Jarrod Gilbert is a sociologist at the University of Canterbury, New Zealand and the lead researcher at Independent Research Solutions.
9  Dr Elizabeth Farrelly is a Sydney-based author, architecture critic, essayist, columnist and speaker
10  *The Sydney Morning Herald*, 11 June 2016.

English physicist Professor Brian Cox. He says ignoring the best evidence is entirely wrong, and is the road back to the cave.

*We think nature's a toy and we're the big kids in the sandpit now, making the rules. I mean come on. Join a few dots here. Last month atmospheric $CO_2$ passed the 400 ppm milestone at a remote Tasmanian monitoring station. It was autumn, but we were still in the longest, hottest summer on record. Tasmania's world heritage forests burnt for the first time in history and UNESCO reported the Great Barrier Reef is 93 per cent bleached, 50 per cent dead.*

You might think that changing weather patterns that affect food production; rising sea levels that contaminate water supplies; increase risks of disastrous flooding, and the prospects of more furious wildfires would be reasons enough to take decisive action, but formulation and implementation of action has been slow.

Yet a further example, the Arctic ice melt is creating a 'new normal' in the far northern marine ecosystem that has created conditions for a whale population boom. In an article published in The Royal

Society's *Biology Letters*[11], biological oceanographer Sue Moore from the US Pacific Marine Environmental Laboratory said that the loss of sea ice in late northern summer means that ocean habitat for whales has expanded. Dr Moore notes that the region north of the Bering Strait, between Alaska and Russia, has lost 75 per cent sea ice by volume and 50 per cent in late summer surface cover and that this has meant an extension of the open-water period in the region by four to six weeks a year, plus a population increase of whales.

She asserts that the growth in whale numbers will also help stabilise the changed ecosystem.

> *Baleen whales act as ecosystem engineers and their recovering may actually buffer the marine ecosystem from destabilising stress associated with rapid change.*

On first pass this might sound like good news. However, this is far from the case as many other marine mammals are suffering from the loss of Arctic ice. Polar bears, walruses and ice seals appear to be particularly vulnerable because they rely on sea ice as a platform for key life functions, such as birthing, nursing young, hunting and resting.

A warming atmosphere also contributes to the spread of pests and diseases once limited to the tropics. It also has detrimental effects on species diversity and other environmental values.

---

This is all 'big picture' stuff I hear you say – well it's not really. Let me try and make it personal. Do you know that coffee is the second largest traded commodity internationally after oil and that coffee production could be cut by up to 50 per cent in a few decades because of the effects of climate change[12]?

---

11  *Is it 'boom times' for baleen whales in the Pacific Arctic region?*, 6 September 2016, Sue E. Moore, Published 6 September 2016.The Royal Society, Biology Letters, London, UK.

12  Climate change could cut coffee production up to 50pc by 2050. Australian Broadcasting Corporation 2016, Josephine Asher http://www.abc.net.au/news/2016-08-29/climate-change-could-halve-coffee-production-by-2050-study-says/7793752

The coffee industry is worth $19 billion worldwide, with more than 2.25 billion cups of coffee consumed every day. Nearly half of all Australians drinking coffee regularly – that may include you. The report indicates that unless action was taken, the effects of climate change would result in coffee supply shortages and increased prices.

"Small changes in the climate could dramatically affect farmers' coffee crops. We're fearful that by 2050, we might see as much as a 50 per cent decline in productivity and production of coffee around the world, which is not so good," said the chief executive of Fairtrade Australia.

"Our concern is primarily for the 25 million farmers out there whose entire livelihoods depend on this incredibly important global commodity," she said.

The chief executive of the Australian Climate Institute John Connor said there was a range of ways climate change was impacting coffee." I guess it's also the story in other agricultural produce as well, but in particular for coffee."

"It's not just the heat, which is a big factor driving some of the regions where coffee is produced uphill. We're also seeing extra diseases increasing and being able to go up into those areas."

Connor said the cost and quality of coffee was likely to pay the price of inaction on climate change.

"For Australians wanting to drink their coffee, it's going to choke up supply lines – we could see higher prices as well as challenges of getting the coffee itself. Flavour and quality is likely to be impacted as well," he said.

The declining supply of popular Arabica coffee beans – grown in East and Central Africa, South America, India and Indonesia – is being felt in the pockets of suburban supermarket shoppers and city cafés. Brands like Maxwell House, Yuban, and Folgers have increased the retail prices of many grinds by 25 per cent or more

Small changes in the climate could dramatically affect farmers' coffee crops and the price of the morning coffee stop.

between 2010 and 2011, in the light of tight supply and higher wholesale prices.

So this brings it right down to the breakfast table or the mid-morning coffee stop, meaning that climate change is personal – the crisis is real and it is upon all of us.

The Union of Concerned Scientists[13] summed up the climate change – coffee dilemma this way:

> *If you're one of those people who needs a cup of coffee to get going in the morning, your world may be changing. In fact, it already is. Climate change is threatening coffee crops in virtually every major coffee producing region of the world. Higher temperatures, intense rainfall coupled with long droughts, and more pests and disease – all associated with climate change – have reduced coffee supplies dramatically in recent years.*

Certainly by now the alarm bells should be ringing loud and long. While this is very serious, the good news is that the most dangerous impacts of climate change may still be avoided if we can move our

---

13  See http://www.ucsusa.org/global_warming/science_and_impacts/impacts/impacts-of-climate-on-coffee.html#.V8kPIM5OKUk

Setting the scene  11

Another traffic jam in Los Angeles. The most dangerous impacts of climate change may still be avoided if fossil fuel-based energy systems are moved towards renewable sources.

fossil fuel-based energy systems towards renewable sources, and increase the use of renewable materials, such as wood.

Some more good news – we already have a fair bit of relevant technology, so a transition in the direction of a carbon-neutral economy is certainly possible. A lack of political will and deliberate confusion generated by special interest groups have contributed to delays in advancing forward more quickly.

You may be surprised to learn that presently 85 per cent of the energy presently generated to run our daily lives comes from fossil fuels. We will consider this issue in more detail in the pages ahead, but in short, climate scientists are warning us about the urgent need

Wood chips from wood processing residues. Ultimately renewable sources of energy, including biomass, will need to provide a substantial proportion of future energy needs.

to halt then reduce the concentrations of atmospheric pollutants or greenhouse gases[14] within a time frame of roughly the next 100 years. I think it is fair to say that such an objective is at present not consistent with our existing vision of economic progress – based on a steady increase in global consumption and growth.

However, more broadly I think that the global community shares a common concern about adverse climate change, and is seeking ways in which the problem may be slowed and eventually overcome. The most obvious way to reduce the atmospheric levels of the major atmospheric pollutant, the greenhouse gas carbon dioxide is to reduce the use of fossil fuels.

Ultimately renewable sources of energy, including biomass[15]

---

14  We will get to greenhouse gases in a fair bit of detail later, but just to start, greenhouse gases are a natural part of the atmosphere. They include water vapour, carbon dioxide, methane, nitrous oxide and some artificial chemicals, such as chlorofluorocarbons. Greenhouse gases absorb and reflect the sun's heat to maintain the Earth's surface temperature.
     The concentration of water vapour, the most abundant greenhouse gas is highly variable. The concentrations of the other greenhouse gases are influenced by human activities, particularly burning fossil fuels (coal, oil and natural gas), agricultural activity and land clearing. Once released into the atmosphere greenhouse gases remain there for a long time increasing the concentrations of gases that trap heat.
     Human activities have increased the level of carbon dioxide and other greenhouse gases. This has increased what is call the 'Greenhouse Effect', trapping more heat in the atmosphere and causing global temperatures to rise. We will also talk about the Greenhouse Effect in more detail later.

15  Other forms of alternative energy include nuclear, wind, solar, hydro and wave.

will need to provide a substantial proportion of our future energy consumption needs. The more frugal use of fossil fuels through more energy efficient vehicles, better public transport networks and the use of less polluting industrial processes will also need to be part of the mix.

---

Other valuable actions for reducing energy consumption, and as a consequence, lowering carbon dioxide emissions, include utilising materials that use less energy in their manufacture and that also store carbon. The use of high energy materials, including steel, concrete and aluminium will need to be restricted. The production of a tonne of steel releases around two tonnes of carbon dioxide into the atmosphere. While on the other hand wood products require only low energy inputs into their manufacture. Also wood products store more than 15 times the amount of carbon dioxide released during their manufacture, but only negligible amounts of carbon are stored in steel, concrete and aluminium products. We will be returning to this important wood 'weapon' in the fight against climate change later.

So wood is our only renewable building and construction material. It has been estimated that a timber beam of one cubic metre stores close to one tonne of carbon dioxide. Compare that to steel, concrete or plastics. Not only do their manufacturing processes require large quantities of electricity, but instead of storing carbon, steel and concrete emit it. For the equivalent one cubic metre beam, concrete releases two tonnes of emissions and steel even more. By the time a concrete skyscraper has been erected, it has produced tens of thousands of tonnes of carbon dioxide.[16] It has been estimated that as a whole, the concrete industry has five times the carbon footprint of the world's airline industry.[17]

---

16  From now on for easy of reading I will refer to carbon dioxide by its chemical formula $CO_2$

17  *New wood: how it will change our skyline. Timber buildings are reaching towards the skies, thanks to breakthroughs in super-strong wood*, 27 August 2016 Greg Callaghan, *Sydney Morning Herald* Good Weekend Magazine.

Wood the only renewable building material. A one cubic metre timber beam stores close to one tonne of carbon dioxide. For the equivalent one cubic metre beam, concrete releases two tonnes of emissions and steel even more.

Hopefully from this short attempt to set the scene, you will get a sense of where this book is heading. While we are going to talk about human-induced climate change and attempt to explain the whys and wherefores, I will spend time commenting on plant photosynthesis – the process at the centre of this issue. In particular, we will discuss the role plants, especially trees and forests, plus wood products can play in assisting to tackle climate change, and in contributing to a sustainable energy and carbon neutral future.

---

I hope I have impressed upon you the urgency of affirmative action to address climate change. Just to underscore this before the end of this chapter, it is worth citing commentary from Deborah Snow and Matt Wade in a major feature, *Back at the Office The newly elected prime minister will have a very long to-do list* in *The Sydney Morning Herald* on 2 July 2016 on the key matters in need of urgent attention by an incoming government. They include taking action on climate change:

> *Record global temperatures for the first five months of 2016 have bleached corals around the world, killing off about 22 per cent of the Great Barrier Reef in one fell swoop.*
> 
> *The events, coming just months after the Paris climate summit agreed to limit warming to 1.5-2 degrees, make avoiding dangerous climate change a priority for Australia whoever wins on Saturday.*

These few pages are perhaps a somewhat pessimistic note to open on, so I hope you will brace yourself for the bumps ahead and read on. We most certainly need to understand that doing nothing will essentially be a death sentence for human society as we know it, and also for the myriad of other life forms inhabiting the planet. So we need to solve the climate change challenge. I have no doubt that to some degree it will be about attempting to push things uphill with a pointed stick, but we have a moral responsibility to try our very best to do so.

CHAPTER 2

# Weather, but not as we know it

And you thought a few thousand Syrian refugees created a drama

A warning to start this chapter, it might sound a bit like the Doomsday Book[18], but we do need to appreciate the gravity of the climate change threat to life on the planet before we launch into the chapters ahead.

Farmer tending her animals in the Oromo region of Ethiopia. While climate change will affect all countries, developing nations are the most vulnerable. Many of these countries depend on climate sensitive activities like agriculture.

18  The Domesday Book was commissioned in December 1085 by William the Conqueror, who invaded England in 1066. The first draft was completed in August 1086 and is a detailed survey of the land held by William the Conqueror. The original Domesday Book has survived over 900 years of English history and is currently housed in a specially made chest at The National Archives in Kew, London.

Clearly the message from science is that unless we bring climate change under control it will lead to catastrophic events, like continuing rising sea levels, food and water shortages and civil unrest. While climate change will affect all countries, developing nations are the most vulnerable. To a large degree these countries depend on climate sensitive activities like agriculture and have only limited capacity to adapt to the consequences of future adverse weather patterns.

Leading Australian academic, Robert Manne[19] expressed the urgency this way:

> ... a re-imagining of the relations between humans and the Earth, a re-imagining that will be centred on a recognition of the dreadful and perhaps now irreversible damage that has been wrought to our common home by the hubristic idea at the very centre of the modern world – man's assertion of his mastery over nature.

So getting to the details of atmospheric pollution – carbon dioxide emissions produced by human activity amounted to about 25 billion tonnes per year for the past decade, or around four tonnes per person.[20] Climate scientists point to the necessity of halving this rate by the middle of this century if already apparent climate change trends are to be kept within tolerable limits.

Further, at least half of all carbon dioxide emissions can fairly be attributed to industrial countries. At 12.6 tonnes per person, carbon dioxide emissions in industrialised countries are five to six times higher than in the developing countries. Across developing countries the average emissions per person is just 2.3 tonnes with a large range of variation from less than a tonne in the poorest countries to

---

19 Robert Manne is Emeritus Professor and Vice-Chancellor's Fellow at La Trobe University in Melbourne. He has twice been voted Australia's leading public intellectual. This quote is from an essay in the December 2015 – January 2016 edition of the Monthly magazine titled: *Diabolical – Why have we failed to address climate change?*

20 *The Carbon Cycle and Atmospheric Carbon Dioxide* 2001 I.C. Prentice et.al
www.grida.no/climate/ipcc_tar/wg1/pdf/tar-03.pdf

Rising sea levels will impact unevenly across the planet. At least 100 million people live below about a metre above sea level, the possible level of sea level rise in the foreseeable future.

4.5 tonnes in those countries with increasing personal income levels. The spectrum is also wide in industrialised countries, from 5.5 tonnes in Malta and Sweden to 20 tonnes in the United States of America. The level of emissions per person in the United States of America is a staggering 200 times more than in some countries of central Africa.[21]

---

The potential of global warming, coupled with sea level rise creates real risks to the longer-term survival prospects for many. Rising sea levels will impact unevenly across the planet and may first push economically weak regions and inhabitants to the limits of survival – or beyond. At least 100 million people live below about a metre above sea level, which according to climate experts, is the

---

21 Some of the detail about climate change impacts in this chapter are taken from Chapter 9 of *The work of nature: how the diversity of life sustains us*, written by Yvonne Baskin in 1997, Island Press, Washington DC, USA. Yvonne Baskin is a widely published science journalist.

Turkish riot police block the highway as Syrian refugees attempt to get to the Bulgarian border, near Edirne, Turkey. The mass movement of Syrian refugees across Europe in 2015 and 2016 caused substantial political, economic and social turbulence.

possible level of sea level rise in the foreseeable future. What will happen to these people? Where will they go? Who will offer them a new home?

The mass movement of Syrian refugees across Europe in 2015 and 2016 caused substantial political, economic and social turbulence across Europe and beyond.[22] Tragic as this refugee crisis is the numbers of people involved were tiny compared with the potential mass movement that could arise from the adverse impacts of climate change.

---

The evidence is that climate change trends are expected to increase the intensity and frequency of extreme weather events, such as storms, floods, droughts and heat waves. Water is already scarce in many regions of the world, with a number of developing countries

---

22  For additional details go to: http://syrianrefugees.eu/

20  By the light of the Sun

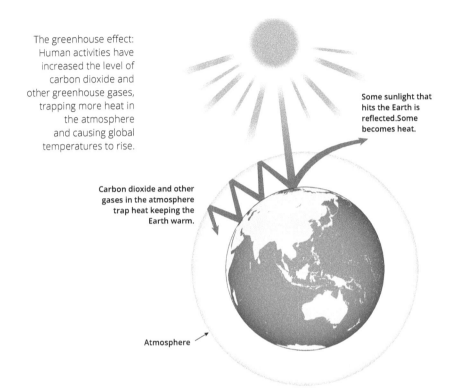

The greenhouse effect: Human activities have increased the level of carbon dioxide and other greenhouse gases, trapping more heat in the atmosphere and causing global temperatures to rise.

Some sunlight that hits the Earth is reflected. Some becomes heat.

Carbon dioxide and other gases in the atmosphere trap heat keeping the Earth warm.

Atmosphere

already not having access to adequate clean water supplies. If, as some climate modelling suggests, temperatures increase by 2.5 degrees Celsius above pre-industrial levels, that equates to 1.7 degrees Celsius above present levels, an additional perhaps three billion people are likely to suffer from water scarcity.[23]

Rising temperatures are also likely to mean that diseases, like malaria, yellow fever, dengue fever and encephalitis currently confined to tropical regions will spread because the area where climatic conditions are suitable for the mosquitoes, ticks and other insects that carry diseases is likely to expand.

---

23  Rising sea levels could lead to regional conflicts, famines and the movement of refugees as food, water and energy resources become scarce. 'Climate change refugees' could be driven from their homes, and will need to be accommodated elsewhere.

Quite simply we must reduce emissions of greenhouse gases into the atmosphere. We should note, without being too despondent, that some greenhouse gases are long-lived, meaning that they hang around in the atmosphere for decades. So even if we take strong action today, temperatures will continue to rise in the immediate future. But if no action is taken temperatures will increase even more. So the sooner we do something the quicker we will be able to take control of our destiny and safeguard the beauty and diversity of our place in the galaxy for our grandkids.

We know now that reducing greenhouse gas emissions will require substantial changes to how we produce and use energy. Studies suggest that the price of doing nothing would be much higher because of the damage and human suffering that unconstrained climate change would unquestionably cause. Citing Lord Nicolas Stern[24]

Lord Stern asserts that moving to a 'low-carbon' economy offers tremendous opportunities for innovation and economic growth.

---

24  Lord Stern is a British economist and academic. He is Professor of Economics and Government Chairman of the Grantham Research Institute on Climate Change and the Environment and Head of the India Observatory at the London School of Economics. He is President of the British Academy (from July 2013), and was elected Fellow of the Royal Society (June 2014).
   Lord Stern was Chief Economist of the European Bank for Reconstruction and Development, 1994-1999, and Chief Economist and Senior Vice President at the World Bank, 2000-2003. He was knighted for services to economics in 2004 and made a life peer as Baron Stern of Brentford in 2007. He has published more than 15 books, his most recent being Why are We Waiting? The Logic, Urgency and Promise of Tackling Climate Change.

in their booklet; *Climate change – what is it all about?*[25] the European Commission say we simply cannot afford just to do nothing, and that climate change will take an increasingly heavy toll on the world economy if we fail to stop it. But on a positive note they quote Lord Stern's work saying:

> ... climate change will cost at least 5 per cent of the world's economic output – or gross domestic product (GDP) – each year, and this could rise to as much as 20 per cent or more in the long-term. The economic impact would be similar to a world war or the Great Depression, the global economic crisis of the 1930s. On the other hand the measures needed to bring climate change under control will cost only one per cent of GDP.

Lord Stern asserts that moving to a 'low-carbon' economy offers tremendous opportunities for innovation and economic growth. Companies quick to develop these technologies he suggests will have a competitive advantage as global demand for these technologies grows.

He confirms that climate-friendly technologies, such as renewable energy sources, already exist while others are close to being ready for widespread use. The development of climate-friendly technologies he believes will create employment opportunities and open up new markets.

But on a pessimistic note science tells us that even if decisive action was initiated to cut greenhouse gas emissions many of the changes that are already underway simply cannot be halted. So like it or not we need to learn to live with climate change. This means determining the probable effects of a changing climate and taking action to minimise impacts, or as Lord Stern says; "By acting now we can save money and effort later."

For example, in some developing countries crops will need

---

25  Climate change – what is it all about? 2009 European Commission, Luxembourg.

to be developed that need less water, or that can tolerate rising temperatures. For already drought-prone countries like Australia, technologies to use water more frugally and wildfire mitigation will need greater science input and resourcing. Housing and building construction will need to place increased emphasis on sustainability and carbon storage credentials of building materials, building storm resistance and energy efficiency. More about this later.

---

Let's recap for a moment. There is no longer any doubt that the climate is changing, or that this change is attributable to human activities, notably the burning of fossil fuels. According to the United Nations Intergovernmental Panel on Climate Change (IPCC)[26] human-related activities contribute six billion tonnes of carbon emissions annually and growing by half a per cent a year. The most optimistic estimates indicate that the concentration of carbon dioxide in the atmosphere will double by 2100.[27]

The first effects of adverse climate change have already occurred, and point the way to much more widespread and destructive changes in the future. According to research detrimental climate change impacts include the reality that the North Pole ice cap is melting and that between 1950 and 2000 its surface area has diminished by 20 per cent.[28] Evidence also points to examples of snow cover diminishing and glaciers retreating.

Sea levels have risen in the 20th century. There is also a significant

---

26  The Intergovernmental Panel on Climate Change (IPCC) is the leading international body for the assessment of climate change. It was established by the United Nations Environment Programme and the World Meteorological Organization in 1988 to provide a scientific view on the current state of knowledge in climate change and its potential environmental and socio-economic impacts. Membership of the IPCC is open to all member countries of the United Nations and World Meteorological Organization. Currently 195 countries are members of the IPCC.

27  Intergovernmental Panel on Climate Change, 2000, IPCC Assessment Report.

28  *Impacts of warning Arctic*, 2005, Arctic Climate Impact Assessment

Melting glacier in Antarctic. Evidence points to global examples of snow cover diminishing and glaciers retreating.

increase in the frequency and severity of natural disasters, such as hurricanes, droughts, earthquakes and floods. Globally the increase in flood damage over recent decades has been a real eye-opener. For instance in the 1960s about seven million people were affected by flooding each year, today that figure stands at 150 million a year.[29]

According to the seminal accounts of Tim Flannery[30], over the last two hundred years, the concentration of carbon dioxide in the atmosphere has increased by a huge 30 per cent. This increase makes earlier human impacts on Earth's atmosphere look trivial. Flannery explains that the excess carbon has come from two sources

---

29 *Managing forests for adaption to climate change*, 2003 Rakonczay, Jr. Z. In ECE/FAO seminar: Strategies for the sound use of wood, Poiana Brasov, Romania.

30 Tim Flannery is an Australian scientist, environmentalist and global warming activist. He was Australian of the Year in 2007; the former Chief Commissioner of the Australian Climate Commission and held the Chair in Environmental Sustainability at Macquarie University in Sydney, Australia. Among his widely acclaimed books on climate change are *The Weather Makers: the History and Future Impact of Climate Change*, 2005 and *Here on Earth: an argument for hope*, 2010. Both published by Text Publishing, Melbourne, Australia.

Tim Flannery, an Australian scientist, environmentalist and global warming activist, asserts that over the last two hundred years, the concentration of carbon dioxide in the atmosphere has increased by 30 per cent.

– around 40 per cent from the destruction of forests and soils, and the rest from mining and burning the fossilised carbon in the form of coal, oil and gas.

The carbon concentration in the atmosphere continues to grow at an ever faster rate. Two hundred years ago the atmospheric concentration of carbon as carbon dioxide was around 2.8 parts per ten thousand. Today it's 3.9 parts – a level not seen for at least three million years – and even if we ceased burning fossil fuels today, it would take several centuries for life, the oceans and Earth's crust to re-absorb the excess. But that will not happen. Instead, if we do nothing, we are on track to increase the concentration to at least seven parts per ten thousand by the end of the century.[31]

Flannery explains that the oceans are a kind of global

---

31  *The Weather Makers: the History and Future Impact of Climate Change*, 2005, Tim Flannery Chapter 3, Text Publishing, Melbourne, Australia.

'thermometer'. The water in oceans expands as heat trapped by the atmosphere is absorbed. Additional water is added as a result of the warming atmosphere from melting glaciers and icecaps. Furthermore, because the oceans are very large, they have a lot of inertia and are slow to respond. By contrast the atmosphere responds quickly to temperature-changing factors, so there can be considerable year-to-year variation.

So the average temperature of the ocean's surface is a more steady guide to global warming trends. It's concerning to discover that sea-level rise is tracking the upper limit of the IPCC's 2007[32] projections of more than a metre within ninety years. Such a rise will be a catastrophe for much of Asia, the east coast of the United States of America and low-lying parts of parts of Europe. Tim Flannery rings the alarm bell this way[33]:

> The incidence of extreme weather events, the decay of the Arctic and Greenland icecaps, and the acidification of the oceans (caused by the absorption of $CO_2$ in sea water) are all proceeding at unanticipated rates. Indeed, the situation is now so severe that the report[34] notes that global average surface temperature is unlikely to drop in the first thousand years after greenhouse gas emissions are cut to zero.

For the first two hundred years following the industrial revolution, emissions increased at an average of 2 per cent per year, but since 2000 they have increased by an average of 3.4 per cent per year, a rate that is driving atmospheric carbon dioxide levels beyond pessimistic predictions. By destroying natural forests across the planet over the last three centuries we have released between

---

32 *Climate Change 2007: Synthesis Report*, Intergovernmental Panel on Climate Change 2007. See: https://www.ipcc.ch/pdf/assessment-report/ar4/syr/ar4_syr.pdf
33 *Here on earth: an argument for hope*. Tim Flannery, 2010 Chapter 15, Pages 197-198. Text Publishing, Melbourne, Australia.
34 *Synthesis Report from Climate Change: Global Risks, Challenges & Decisions*, Richardson, K. et al. 2009, University of Copenhagen, Denmark.

200 and 250 billion tonnes of carbon into the atmosphere making up between 22 and 43 per cent of all the carbon released during that period. Again Flannery puts the situation clearly and somewhat chillingly[35]:

> Climate science is now so well advanced that we can anticipate the kind of event that may, if we do not reduce the stream of greenhouse gas pollution, initiate the end of the great 'us' that is our global civilisation. With no warning a gargantuan ice sheet will began to collapse. It will mark the beginning of an irreversible process and even if the initial rise in sea level it causes is just a few centimetres, it will herald the abandonment of our coasts.

---

Some of this is bleak for sure. However, I am attempting to be a realist not an alarmist. Spelling out the facts will help us squarely confront climate change challenges and fashion workable solutions. Having said that, in the next chapter I want to continue to outline some real and present dangers as a consequence of imminent changes to weather patterns impacting on our planet.

---

35 *Here on Earth: an argument for hope.* Tim Flannery, 2010 Chapter 23, Page 273. Text Publishing, Melbourne, Australia.

CHAPTER 3

# People by the billion

Population explosion put pressure
on the planet

Let's be blunt here – it's people, or more the excess number of people – that are at the centre of the climate change challenge. A finite planet and too many people – and numbers are growing.

David Suzuki says humanity has become a force like no other species in the 3.8 billion years of life's existence on Earth and this has occurred with explosive speed. He writes that it took all of human existence to reach a population of one billion early in the nineteenth century. Since then, in less than two centuries, it has shot past seven billion.

> *Each time the population doubles, the number of people alive is greater than the sum of all other people who have ever lived, but now we are also living more than as long as people did in the past.*
>
> *We are the most numerous mammal on the planet, and our numbers and longevity alone mean that our ecological footprint is huge: it takes a lot of air, land and water to meet our basis needs.*[36]

The impacts of growing humanity are bad news for climate change mitigation. Population expansion will need to be managed, requiring government commitment and political courage.

---

World population grew to over seven billion in mid-2012 after having passed the seven billion mark in 2011. Developing countries accounted for 97 per cent of this growth due to the dual effects of

---

[36] The Legacy: An Elder's Vision for our Sustainably Future, David Suzuki, 2013 Page 15 Allen & Unwin, Vancouver, BC, Canada.

Women devotees participate in a festival walk through the streets in Attukal, Trivandrum, India. The impacts of growing humanity are bad news for climate change mitigation. Population expansion will need to be managed, requiring government commitment and political courage.

high birth rates and young populations. Conversely, in developed countries the annual number of births barely exceeds deaths because of low birth rates and much older populations. By 2025, it is likely that deaths will exceed births in the developed countries, the first time this will have happened in history.[37] The decline in population growth in developed countries is in large measure the consequence of near universal reduction of fertility. Women are marrying later or not at all, postponing childbearing and having fewer or no children.

While virtually all future population growth will be in developing countries, the poorest of these countries will see the greatest percentage increase. Such countries have especially low incomes,

---

37  Population Reference Bureau *Carl Haub*, July 2012 See: www.prb.org/Publications/Datasheets/2012/world-population-data-sheet/fact-sheet-world-population.aspx

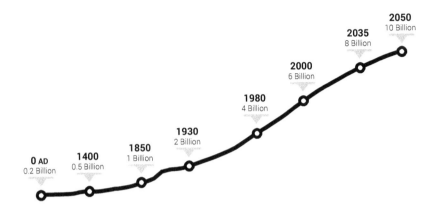

World population grew to over seven billion in mid-2012 after having passed the seven billion mark in 2011. Developing countries accounted for 97 per cent of this growth due to the dual effects of high birth rates and young populations. (Source: United Nations Population Division).

high economic vulnerability, and poor human development indicators such as low life expectancy, very low per capita income, and low levels of education. Of these countries, 33 are in sub-Saharan Africa, such as Burundi, Ethiopia, Mozambique, and Zambia; 14 in Asia, including Bangladesh, Cambodia, Nepal, and Yemen; and one in the Caribbean, Haiti. They are growing at 2.4 per cent per year and are projected to at least double to 2.3 billion by 2050.[38]

With a current population of 4.3 billion, Asia will most likely experience a much smaller proportional increase than Africa, but will still add about a further billion people by 2050. Much of Asia's future population growth will be determined by what happens in China and India, two countries that account for about 60 per cent of the region's population.

---

38  World population outlook to 2300 2012 United Nations Department of Economic and Social Affairs Population Division, New York. See www.un.org/esa/population/publications/longrange2/WorldPop2300final.pdf

I want to convey something of the flavour of the human population across Southeast Asia as an example of the growing population division between East and West.

The nineteenth and twentieth centuries have seen an extraordinary multiplication of the population right across Southeast Asia, from little more than 30 million in 1800 to 80 million in 1900, and around 525 million in 2000. A rapid decline in infant mortality and longer life expectancy from the late 1940s, plus a rise in fertility in some countries have contributed to an acceleration in population growth from the 1950s. However, to put this into context, even the recent decades of rapid population increase have left overall population densities in Southeast Asia below those of countries such as Japan, Korea, Bangladesh and India.

Crowded Brazilian Rocinha favela shanty town in Rio de Janeiro Brazil. There is a radically different demographic situation between developed and developing countries.

However, in a further example of the escalating influence and ascendancy of the East, there has, over the course of the last century, been a dramatic reversal in the demographic balance between Southeast Asia and Europe. Take Indonesia as an example; at the start of the twenty first century, the Indonesian population exceeded that of Russia – the largest European country – by almost 100 million. There are now more Vietnamese and Filipinos than Germans. Thailand, a medium-sized Southeast Asian country has a larger population than either Italy or the United Kingdom.

Interestingly, within Asia, several of the more economically advanced countries such as Japan, Singapore, South Korea, and Taiwan have low birth rates. In Japan, 24 per cent of the population is already aged 65 and older, a proportion certain to continue growing.

---

Latin America and the Caribbean is the developing region with the smallest proportional growth expected by 2050, from 599 million to 740 million, largely due to fertility declines in several of its largest countries such as Brazil and Mexico.

The very sharp decline in fertility in developed countries is a key aspect of current global population dynamics. Yet not all developed countries tell the same story. In countries such as France and Norway, social programs to support families, such as generous maternity leave and subsidies for child care have kept birth rates stable.

However, overall, Europe is likely to be the first region in history to see long-term population decline largely as a result of low fertility in Eastern Europe and Russia. Europe's population is projected to decrease from 740 million to 732 million by 2050. The population of the 27 countries in the European Union, around 502 million, should roughly maintain their current size, even with large increases in the elderly population compared with younger age groups.

In Australia, Canada, New Zealand, and the United States of America, continued growth from higher births or continuing immigration, or both, are expected, although these countries have not been immune to lower birth rate trends.

•

The radically different demographic situation between developed and developing countries illustrates the "demographic divide"[39] – the vast gulf in birth and death rates among the world's countries. On one side of this divide are mostly poor countries with relatively high birth rates and low life expectancies. On the other side are mostly wealthy countries with birth rates so low that population decline is all but guaranteed and where average life expectancy extends past age 75, creating rapidly aging populations.

Although growth rates will fall, the annual increase in world population will remain – 57 million a year on average between 2000 and 2050. This is smaller than the 71 million people added annually between 1950 and 2000, but still substantial. It means that, on average each year for 50 years, world population will expand by about as many people as now live in Italy. The increase, over 50 years, will be more than twice the current population of China, or more than twice the current population of all more developed regions combined. Although population growth will eventually subside, and a variety of countries will see little or no

---

39  The "demographic divide" is the large gap in birth and death rates among the world's countries. On one side are mostly poor countries with relatively high birth rates and low life expectancies. On the other side are mostly wealthy countries with birth rates so low that population decline is all but guaranteed and where average life expectancy extends past age 75, creating rapidly aging populations. The "demographic divide" also involves a set of demographic forces that will affect the economic, social, and political circumstances in these countries and, consequently, their place on the world stage. For further information see: www.prb.org/Publications/Articles/2005/TheDemographicDivideWhatItIsandWhyItMatters.aspx

The vast slums of Juba, the capital of South Sudan. Population trends are *the elephant in the room* when it comes to reasons behind escalating adverse climate change.

population growth, for the world as a whole the next 50 years can hardly be characterized as demographically tranquil.

United Nation's population data[40] confirm the continuation of long-term global demographic trends and a larger global population than previously projected. Nations cannot afford to ignore trends, including reduced fertility among most nations and limited family planning for others, plus growing ranks of the elderly and accelerated urbanization pose significant challenges. Life expectancy is also higher in most countries.

Population problems indicate the pervasive and depressive effect that uncontrolled growth of population can have on many aspects of human welfare. Nearly all our economic, social, and political

---

40  *World population outlook to 2300* 2012 United Nations Department of Economic and Social Affairs Population Division, New York. See www.un.org/esa/population/publications/longrange2/WorldPop2300final.pdf

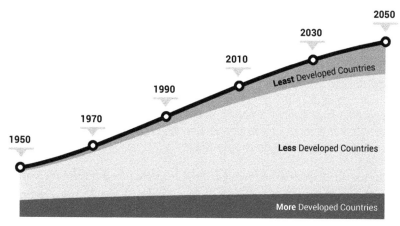

Data confirms a continuation of long-term global demographic trends and a larger global population. Nearly all economic, social, and political problems become more difficult to solve in the face of uncontrolled population growth. (Source: *World population outlook*, United Nations Department of Economic and Social Affairs Population Division)

problems become more difficult to solve in the face of uncontrolled population growth. It is clear that even in wealthier nations many individuals and families experience misery and unhappiness because of the birth of unwanted children. The desirability of limiting family size is now generally, though not universally, recognized, particularly among the better-educated segments of the population in many countries.

According to the *Science Summit on World Population: A Joint Statement by 58 of the World's Scientific Academies*[41] it took hundreds of thousands of years for humanity to reach a population level of ten million, only 10,000 years ago. This number grew to 100 million about 2000 years ago, and to 2.5 billion by 1950. Within less than the span of a single lifetime, it has more than doubled to 5.5 billion in 1993. We face the prospect of a further doubling of the population within the next half century.

---

41  As detailed in *The Legacy: An Elder's Vision for our Sustainably Future*, David Suzuki, Allen & Unwin, Vancouver, BC, Canada. Pages 20-21

David Suzuki sums up the connection between population increase and rapid technological innovation[42].

> *The rapid increase in human numbers has coincided with accelerating technological innovation, which has risen even more steeply over the same period as population growth. Technology has enabled us to keep up with the escalating demands on the planet from more and more people. But greater numbers of people plus consumptive demand have placed a huge burden on the biosphere.*

So population trends are what might be described as *the elephant in the room* when it comes to reasons behind escalating adverse climate change. How to address this critical issue will be a core issue if humanity is going to continue to survive on planet Earth.

Now moving on, in the next chapter we are going to scrutinize more closely central features of the climate change-related phenomena.

---

42   *The Legacy: An Elder's Vision for our Sustainably Future*, David Suzuki, Pages 20-21, Allen & Unwin, Vancouver, BC, Canada.

CHAPTER 4

# Feeling the heat

The need to tackle climate change has become urgent

A report published by the Australian Climate Council titled: *The Heat Marches On* [43] provides a confronting analysis of emerging climate change induced impacts in Australia and elsewhere. The report cites exceptionally long and hot spells in early March 2016 that contributed to an escalating number of heat records in Australia and across the globe.

The Climate Council report talks about more 40 degrees Celsius days in the 2015-2016 summer in Perth than ever before, and that Sydney experienced a record breaking 39 consecutive days over 26 degrees Celsius, eclipsing the previous record of 19 days.

Records were also broken globally, with January and February 2016 being significantly hotter than any other January and February on record. The Climate Council report says these temperature records are driving dramatic and unprecedented climate impacts.

In Australia record warm sea surface temperatures are threatening the Great Barrier Reef with widespread coral bleaching evident. Prolonged hot temperatures have also contributed to a major algae bloom in the Murray River, and hot and dry conditions over the 2015-16 summer were considered a factor in particularly destructive wildfires in Tasmania.

2016 was Australia's fourth-warmest year on record.[44] The "area-averaged mean temperature" for 2016 was 0.87 degrees Celsius

---
43  *The Heat Marches On* 2016 Steffen and Jacqui Fenwick, Climate Council of Australia.
44  Australia's the national weather observational dataset commenced in 1910.

A bushfire at night: Australia continues to suffer from more frequent and worsening extreme heat events.

above the 1961 to 1990 average. Maximum temperatures were 0.70 degrees Celsius above average, and minimum temperatures were 1.03 degrees Celsius above average. Minimum temperatures were the second warmest on record.[45]

It is also notable that seven of Australia's ten warmest years have occurred since 2005. The cities of Darwin and Sydney had their warmest years on record in 2016 for both maximum and minimum temperatures, whilst Hobart had its warmest nights on record and warmest annual mean temperature. For Brisbane the annual mean temperature was the warmest on record.

The Climate Council report warns that as Australians continue to suffer from more frequent and worsening extreme heat events, the path to tackling climate change is becoming even more urgent.

---

45 Annual climate statement 2016, 5 January 2017 Australia Government, Bureau of Meteorology. See: http://www.bom.gov.au/climate/current/annual/aus/

The Climate Council recommends that no new coal mines be commissioned; that existing coal mines and coal-fired power stations be phased out, and renewable energy projects rapidly scaled up.

The report notes that the United States of America declared a moratorium on new coal mines on federal land, and that as renewable energy ramps up the use of coal fell to record lows in 2015. Also that China has pledged to shut 1000 coal mines in 2016, plus modelling shows their emissions may have already peaked ahead of schedule.[46]

In contrast, Australia's fossil fuel emissions have begun to rise again, particularly in the electricity sector, with electricity emissions increasing by 3 per cent in 2014-2015.

---

February temperatures in 2016 in some American states were substantially above the long-term average, including Alaska which had its warmest February on record at 6.8 degrees Celsius above average. In central England, which has the longest temperature record in the world[47], the 2015-16 winter was the second warmest ever, and only 0.1 degree Celsius behind the previous record set in 1869.

The average global surface temperature – land and ocean combined – for February 2016 was 1.2 degrees Celsius above the twentieth century average, the highest February temperature on record and the highest departure from average on record for any month. The temperatures on land were particularly high, with the average land surface temperature 2.3 degrees Celsius above the twentieth century average, and further above average than any month on record.[48]

---

46 *China's changing economy: implications for its carbon dioxide emissions.* Green F. and Stern N. 2016. Climate Policy, 2016.
47 Climate records in England date back to 1659.
48 For further information see: National Oceanic and Atmospheric Administration (USA) 2016. Global Analysis - February 2016; https://www.ncdc.noaa.gov/sotc/global/201602

Dredge loads the truck with coal. The United States of America declared a moratorium on new coal mines on federal land. Also China has pledged to shut 1000 coal mines.

2015 was the hottest year on record globally, marking the fourth time this century that the annual global temperature record has been broken. Fourteen of the fifteen hottest years on record have occurred in the last fifteen years. Worsening extreme heat events in Australia and around the world take their toll on human health, the environment, agriculture, infrastructure and many other facets of daily life.

———————————— • ————————————

Unusually high sea surface temperatures in the tropics during the 2015-16 summer have, as mentioned, contributed to a major coral bleaching event affecting the Great Barrier Reef. Coral reefs take many years, or even decades to recover after severe bleaching events. Repeating bleaching can cause the death of coral and the conversion of the reef to an algae-dominated ecosystem.[49]

---

49  For further information see: Great Barrier Reef Marine Park Authority, 2016b. *Managing the Reef: Coral bleaching*. 2016; www.gbrmpa.gov.au/managing-the-reef/threats-to-the-reef/climate-change/what-does-this-mean-for-species/corals/what-is-coral-bleaching.

Beyond the catastrophic environmental consequences, the impact of extreme heat on these high-profile ecosystems also comes at an economic cost. The Great Barrier Reef and Tasmania's forests deliver billions of dollars of value-added economic contribution annually.

Arctic sea ice typically reaches its maximum extent for the year in mid to late March. The ice cover on the Arctic Ocean has a major influence on the broader climate system and on regional warming. As the area of summer ice melt expands, more dark ocean water is exposed, which absorbs rather than reflects incoming sunlight and warms the region even further. For this reason, a rapid decline in sea ice extent is of major concern for both the Arctic and the global climate system.

Australia, and the rest of the world, are in uncharted territory when it comes to the warming of the world's climate. February 2016 obliterated the global temperature record, reaching 1.2 degrees Celsius above the long-term average for the first time in February.

Unusually high sea surface temperatures has contributed to a major coral bleaching events. Coral reefs can take decades to recover after a severe bleaching event.

Columbia Glacier, Prince William Sound, Alaska. A rapid decline in sea ice extent is of major concern for both the Arctic and the global climate system.

The Climate Council contends that when put in the context of the long-term global warming trend, the extreme heat events of early 2016 add to the overwhelming evidence for human activity-driven climate change.

---

It may be considered anecdotal, but media reports across Southeast Asia echo the sentiments expressed in the Climate Council report. In April 2016 and the headline on the front page of the Malaysian daily newspaper *The Star* on 5 April read: Heat taking its toll. The opening paragraph was:

*It's heartache for livestock and fish farmers whose animals are showing signs of severe stress due to the scorching sun. They have warned of higher prices in the coming months. Meanwhile, haze has worsened in Sabah while some states have warned of water cuts.*

Inside the paper articles covered weather impacts on fish farming, animal production, water supply scarcity and smoke from uncontrolled forest and grass fires.

Malaysia has the largest marine fish farming industry in Southeast Asia, and according to the Marine Fish Farmers Association the

Climate change is adversely affecting the fish farming industry, with prices climbing because of fish stock losses due to rising sea water temperatures.

weather is adversely affecting the industry, with prices due to climb because of fish stock losses due to rising sea water temperatures. Deputy President of the Marine Fish Farmers Association Mohamed Razalin Mohamed was reported as saying:

Forest fire in Central Greece: Thick smoke from uncontrolled forest fires causes breathing problems and contributes to adverse climate change becoming an environmental and economic reality.

> *Groupers, snappers and barramundi are dying in cages due to the red tide or the growth of algae, which competes with the fish for oxygen. Just last month, 100 tonnes of fish worth RM4[50] million that were destined for both local and overseas markets were wiped out in fish farms in Palau Kukup and Johor.*

Other articles in the newspaper outlined water shortages and that 128 pig farmers were experiencing difficulties because of a combination of a lack of water and hot weather and the need to hose pigs down daily to keep them cool. What was called an "unprecedented" hot dry spell of weather was also leading to reduced appetite and a decrease in birth rates among animals.

---

50 I RM (Malaysian Ringgit) equals about USD 0.25.

In Sabah, it was reported that thick smoke from uncontrolled forest fires was causing breathing problems for people on the country's west coast. Because of these fires children were not attending schools and face mask sales from pharmacies had skyrocketed. The Sabah Education Department deputy director, Maimunah Suhaidul reported that 83 primary and secondary schools with more than 20,000 students would stay closed until atmospheric conditions improved.

Such reports coupled with science-based studies certainly suggest that adverse climate change is becoming an environmental and economic reality right now, reinforcing the urgency of calls for more immediate action to confront the here-and-now impacts on everyday lives of the planet's inhabitants.

CHAPTER 5

# Living in a greenhouse

Things are hotting up fast and natural causes
cannot account for such warning

We all know that the atmosphere acts like a transparent, protective covering around the Earth. It lets in sunlight and retains heat. Without the atmosphere, the Sun's heat would just rebound off the Earth's surface back into space. If that was to happen it would be some 30 degrees Celsius colder here on Earth and everything would freeze solid.

It is important not to confuse the natural greenhouse effect with the contribution we are making to intensify this effect through rapidly increasing carbon dioxide (because of its frequent use from this point forward I will refer to carbond dioxide by it's chemical formula $CO_2$) emissions.

The atmosphere acts like the glass sides of a greenhouse. Hence this is referred to as the *Greenhouse Effect*. It is the greenhouse gases in the atmosphere that we discussed in the previous chapter that trap heat and are responsible for this effect.

Canadian David Suzuki[51], to whom we referred in Chapter 1, explained greenhouse gases in this way:

> Greenhouse gases such as carbon dioxide, methane, and water vapour occur naturally, while more potent molecules like CFCs[52] are synthesized by us. They allow sunlight to penetrate the atmosphere but, like a blanket, reflect infrared (heat) wavelengths back onto Earth's surface.

---

51  *The Legacy: An Elder's Vision for our Sustainably Future*, David Suzuki, 2013 Pages 25-26 Allen & Unwin, Vancouver, BC, Canada.

52  CFC's or chlorofluorocarbons are a group of manufactured chemical compounds that contain chlorine, fluorine, and carbon. They are colourless, odourless, non toxic, and stable when emitted. When they are emitted and reach the upper atmosphere, they break apart and release chlorine, which destroys the earth's ozone layer. CFCs can last for more than 100 years in the atmosphere. Because they destroy the ozone layer, CFCs have been banned from production. CFCs have been widely used as refrgerants, propellants in aerosol applications, and solvents. They have now been replaced with other products such as hydro fluorocarbons (HFCs).

*The continued addition of greenhouse gases to the atmosphere, thereby thickening the blanket effect, is projected to have potentially catastrophic ecological consequences.*

*In nature, everything is connected, so as more heat trapping gases are added to the atmosphere, polar ice sheets begin to melt, ocean waters warm and expand, and terrestrial ecosystems begin to change as animals and plants move in order to remain within their temperature comfort zone.*

---

The scientific evidence is clear. We know that past ice ages have alternated with warmer periods, and that the average temperatures on Earth have varied between about 9 degrees and 22 degrees Celsius.[53] The current average temperature is 15 degrees Celsius although fluctuations have occurred due to natural causes, such as slight variations in the Earth's orbit, changes in the sun's activity and the after effects of major volcanic eruptions.

City smoke pollution: Carbon dioxide emissions from industrial activity contributing to climate change.

---

53  *Climate change – what is it all about?* 2009 European Commission, Luxembourg

It's true that the world as we know it today is the result of the climate remaining stable for the past, say 10,000 years with only small changes of less than one degree Celsius over about each hundred years. These stable conditions have enabled us to exist as we are today – stable communities and lifestyles. But now things are heating up fast and natural causes alone cannot account for such rapid warning, which is unprecedented for at least the past 2000 years. In addition, the concentrations of $CO_2$ in the atmosphere are today the highest they have been for at least 650,000 years.[54]

The principle greenhouse gas produced by human activities is $CO_2$. It makes up about 80 per cent of all the emissions of greenhouse gases generated. $CO_2$ is released when coal, oil and natural gas are burnt, and fossil fuels are still the most common energy source. We burn them to produce electricity and heat, and we use them to fuel our cars, ships, planes and other machinery.

---

An interesting aside, the term *Greenhouse Effect* is really somewhat inaccurate. Sure temperatures increase inside a greenhouse, but $CO_2$ levels are typically very low during daylight hours. Commercial growers often deliberately release $CO_2$ into greenhouses to stimulate growth. Higher $CO_2$ levels equate to high rates of net photosynthesis and typically the growth of tomatoes or other plants being grown. Raised $CO_2$ levels may therefore not be a problem at least in terms of plant productivity.

Since the Industrial Revolution in the eighteenth century, we have been producing greenhouse gases in ever increasing amounts. Since 1850, the average global temperature has increased by 0.76 degrees Celsius.[55]

---

54  *The Weather Makers: the History and Future Impact of Climate Change*, 2005, Tim Flannery The slow awakening Pages 1-8, Text Publishing, Melbourne, Australia.

55  *Managing forests for adaption to climate chang*e 2003 Rakonczay, Jr., Z. 2003 ECE/FAO Seminar: Strategies for the Sound Use of Wood, Romania.

Greenhouse, Birr Castle Gardens, County Offaly, Ireland: Higher $CO_2$ levels equate to high rates of net photosynthesis, that may not be a problem, at least in terms of plant productivity.

This warming trend is due to the growing quantities of greenhouse gases released by human activities and it is accelerating. The rate of temperature increase has risen from 0.1 degree Celsius per decade over the last hundred years to 0.2 degrees Celsius in the last decade. It is predicted that the average global temperature is most likely to increase further between 1.8 degrees and 4.0 degrees Celsius over the course of this century, but could rise by as much as 6 degrees.[56]

Temperature rises of this magnitude may not seem much until it is remembered that during the last Ice Age, which ended 11,500 years ago, the average global temperature was just 5 degrees Celsius cooler than today, yet polar ice covered much of the Northern Hemisphere. So it is very clear that a few degrees makes a lot of difference to the climate.

---

56 *Tackle Climate Change: Use Wood*, 2009 CEI-Bois, www.cei-bois.org

It should be noted that it is rates of change that are dangerous, after all life is flexible, and if given sufficient time, it can adapt to changing conditions. So it is the rate, not the direction or overall scale of change that is important. Climate scientists taking this line argue that warming rates above 0.1 degrees Celsius per decade are likely to rapidly increase the risk of significant ecosystem damage.[57] Similarly, rates of sea level rise above two centimetres per decade would be dangerous, as would a rise of five centimetres overall.

The National Snow and Data Center at the University of Colorado in the United States of America indicates that in the next few years, with more typical warmer conditions, there are likely to be some very dramatic Arctic sea ice losses. The research centre notes that the ten years with the lowest extent of Arctic sea ice have all been within the last ten years, and that the current rate of loss of Arctic summer sea ice of 13 per cent per decade is equivalent to an annual loss greater than the size of Scotland.[58]

The research centre stresses that the long term loss of ice will affect animals including polar bears. Polar bears need sea ice because they hunt their main food – fat-rich seals – exclusively from the surface of the ice. If ice free periods are too long, bears cannot build the energy reserves they need to see out the lean months.

Up to one in five adult males, and much higher numbers of young and old bears will starve every year in these conditions. The lack of food and shelter for pregnant females will see them failing to keep a significant percentage of their cubs alive.

It is estimated that by 2075, losing all newborn cubs could be commonplace in many parts of the Canadian Arctic Archipelago –

---

57  *Rate of Climate Change To Soar By 2020s, With Arctic Warming 1°F Per Decade* 10 March 2015, ThinkProgress at:https://thinkprogress.org/rate-of-climate-change-to-soar-by-2020s-with-arctic-warming-1-f-per-decade-85db70fb9d1

58  *Arctic sea ice maximum at record low for third straight year, National Snow and Data Center, see*: https://nsidc.org/news/newsroom/arctic-sea-ice-maximum-record-low-third-straight-year.

Average monthly Arctic sea ice extent (September) from 1979 through 2016. Dramatic Arctic sea ice losses evident (Source: United States Environmental Protection Agency).

a vast wilderness home to a quarter of the estimated global population of 20,000 polar bears. Without much of a new generation coming through to replace older bears, extinction of the species would seem certain.

But the question of what constitutes dangerous climate change raises another question - dangerous to whom? For the Inuit whose primary food sources of caribou and seal are now difficult to find as a result of climate change. Also an economically and culturally damaging threshold has already been crossed for low-lying islands in the Pacific Ocean.

Climate change is expected to increase the intensity and frequency of extreme weather events, such as storms, floods, droughts and heat waves. Water is already scarce in many regions of the world. Almost one-fifth of the world's population, 1.2 billion people, do not have access to clean drinking water. If global temperatures increase by 2.5 per cent above pre-industrial levels (that is, around

1.7 degrees Celsius above present levels), an additional 2.4 to 3.1 billion people worldwide are likely to suffer from water scarcity.

In his book: *The Weather Makers: the History and Future Impact of Climate Change*, Tim Flannery sums up the situation[59]:

> When we consider the fate of the planet as a whole, we must be under no illusions as to what is at stake. Earth's average is around 15 degrees Celsius, and whether we allow it to rise by a single degree, or 3 degrees Celsius, will decide the fate of hundreds of thousands of species, and most probably billions of people. Never in the history of humanity has there been a cost-benefit analysis that demands greater scrutiny.
>
> We understand that everything in the world was connected, that what we did had repercussions, and that therefore every act was laden with responsibility.

We have already quoted David Suzuki at the start of this chapter, but here is another one that expresses similar sentiments to Tim Flannery, but in more philosophical terms[60]:

> Nature was our touchstone and our reference point and dictated the way we interacted with it. But as economics and politics have increasingly come to dominate our decisions and actions, we have lost our sense of place in the world and our reverence for nature. We need a new relationship with the planet that is, in fact, our ancient understanding.

---

Well, enough of this broad understanding of the Greenhouse Effect and its potential impact on our home in the galaxy for now. In the next couple of chapters we will get down to some of the detail underlying the climate change phenomenon. The logical place to start is at the molecular level – so read on.

---

59 *The Weather Makers: the History and Future Impact of Climate Change*, 2005, Tim Flannery Chapter 17, Page 170, Text Publishing, Melbourne, Australia.
60 *The Legacy: An Elder's Vision for our Sustainably Future*, David Suzuki, 2013 Page 5 Allen & Unwin, Vancouver, BC, Canada.

CHAPTER 6

# Oh no – not a chemistry lesson

Calculating a safe carbon budget for humanity

At this stage in the proceedings it will be useful if we back up a bit to hopefully better understand the intricacies behind climate change. Trust me, in attempting to get our heads around global temperature warming as a result of greenhouse gas build-up it will be helpful to take the time to look at the chemistry at a molecular level. So in this chapter it's back to the chemistry lab for a quick refresher course.

A bit of background to start; life on planet Earth evolved within a thin, fragile envelope of soil, water and air called the biosphere for hundreds of millions of years before our earliest human ancestors climbed down out of the trees. The planet's plant and animal life was cocooned by the biosphere, and the ongoing survival of life forms of every shape and intellect are absolutely dependent on its continuing efficient functioning. Unfortunately, adverse climate change means that this is far from assured.

David Suzuki puts this absolute dependence on the biosphere in blunt terms[61]:

> As biological creatures, we have absolute requirements from the biosphere. Deprived of air for a few minutes, water for a few days, or food for a few weeks, we perish. ... These are immutable realities defined by our biological makeup.

For the last 10,000 years or so the Earth's average surface temperature

---

[61] *The Legacy: An Elder's Vision for our Sustainably Future*, David Suzuki, 2013 Page 5 Allen & Unwin, Vancouver, BC, Canada.

Plant Earth, a thin fragile envelope of soil, water and air. The ongoing survival of life forms of every shape and intellect is absolutely dependent on the continuing efficient functioning of this biosphere.

has been around 14–15 degrees Celsius. This has enabled humanity to develop and establish a stable society and communities – planting crops, domesticating animals, building cities and so on.

The planet's surface temperature control is a complex and delicate mechanism and $CO_2$ plays a critical role in maintaining the balance necessary to all life. Whereas on 'dead' planets such as Venus and Mars, $CO_2$ makes up most of the atmosphere, and it would do so here if plants and Earth's processes did not keep it within bounds. As it is, $CO_2$ makes up around three-to-four parts per 10,000 of the Earth's atmosphere. This might seem to be a modest amount, yet, as Tim Flannery explains[62] it has a disproportionate influence on the planet:

> Because we create $CO_2$ every time we drive a car, cook a meal or turn on the light, and because the gas lasts around a century in the atmosphere, the proportion of $CO_2$ in the air we breathe is increasing.[63]

---

62  *The Weather Makers: the History and Future Impact of Climate Change*, 2005, The slow awakening Page 5, Text Publishing, Melbourne, Australia.

63  *The Weather Makers: the History and Future Impact of Climate Change*, 2005, The slow awakening Page 5, Text Publishing, Melbourne, Australia.

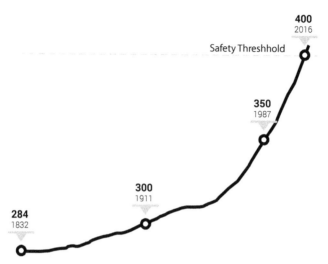

Prior to the start of the Industrial Revolution there were about 280 parts per million of $CO_2$ in the atmosphere. Today the figure has ticked over 400 parts per million (Source: American Chemical Society).

Since the start of the industrial revolution there has been an appreciable increase in greenhouse gas emissions mainly due to $CO_2$ from the burning of fossil fuels, but also from forest destruction. As a result, mean temperatures are expected to rise at a rate of 0.1 to 0.4 degrees Celsius per decade during the first half of this century.[64]

The fact that a known proportion of $CO_2$ exists in the atmosphere allows us to calculate, in round numbers, a carbon budget for humanity. Prior to the start of the Industrial Revolution around 1800, according to Tim Flannery[65] there were about 280 parts per million of $CO_2$ in the atmosphere, which equates to around 586 gigatonnes (billion tonnes) of $CO_2$. These figures relate only to the carbon in the $CO_2$ molecule.

---

64 *Managing forests for adaption to climate change.* ECE/FGAO Seminar: Strategies for the Sound Use of Wood 2003 Poiana Brasov, Romania.

65 *The Weather Makers: the History and Future Impact of Climate Change*, 2005, The slow awakening Chapter 3, Text Publishing, Melbourne, Australia.

Just to pause for a moment here, it is important to appreciate that the difference between a tonne of carbon and a tonne of $CO_2$. The amount of carbon in a molecule of $CO_2$ is 27 per cent. Therefore 3.7 tonnes of $CO_2$ is equivalent to 1 tonne of carbon.

So if the world produces 5500 million tonnes of $CO_2$ per year from fossil fuel combustion, this is equivalent to 1485 million tonnes of carbon. If all this were to be sequestered in trees at 5 tonnes of carbon per year, the amount of forest required would be 297 million hectares. So clearly just on their own, more trees and forest is not a realistic total solution to combatting climate change.

Today the figure has ticked over 400 parts per million or around 830 gigatonnes. If we wish to stabilise $CO_2$ emissions at a level double that which existed before the Industrial Revolution we would have to limit future human emissions to around 600 gigatonnes. Just over half of this would stay in the atmosphere, raising $CO_2$ levels to around 1100 gigatonnes, or 550 parts per million, by 2100.

Sure – the arithmetic is a bit mind-numbing – sorry. Flannery says[66]:

*This, incidentally, would be a tough budget for humanity to abide by, for if we use fossil fuels for only another century, that equates to a budget of 6 gigatonnes per year. Compare this with the average of 13.3 gigatonnes of $CO_2$ that accumulated every year throughout the 1990s (half of this from burning fossil fuel), and the projection that the human population is set to rise mid-century to 9 billion, and you can see the problem.*

---

Now let us take a closer look at the chemistry of carbon in the atmosphere on, and beneath the surface of the planet. Carbon is constantly shifting in and out of our bodies as well as from rocks to sea or soils, and from there to the atmosphere and back again. These

---

66  The Weather Makers: the History and Future Impact of Climate Change, 2005, The slow awakening Chapter 3, Page 29, Text Publishing, Melbourne, Australia.

so called *carbon cycle* movements are extraordinarily complex and are governed by temperature, the availability of other elements and the activities of species such as plants and ourselves.

As most carbon exchanges involve $CO_2$, what are commonly known as *carbon sinks* are really sinks of $CO_2$ – those elements in the cycle able to capture $CO_2$ and to reduce its concentration in the atmosphere.

All living things are carbon sinks, as are oceans, fossil fuel deposits and various rocks. Some of these carbon sinks are very large, but they are not infinite. Over eons much $CO_2$ has been stored in or below Earth's surface. This occurs when dead plants are buried and become fossil fuels. The volume of carbon circulating around our planet is enormous. Around a trillion tonnes of carbon is tied up in living things while the amount buried underground is far, far greater.[67] And for every molecule of $CO_2$ in the atmosphere, there are fifty in the oceans.

It is this buried carbon that allows oxygen to exist in the atmosphere. Should we somehow be able to take all of that fossil carbon and return it to the atmosphere by burning it, we would use up all the oxygen in the atmosphere. Even volcanic eruptions that contain much $CO_2$ can disturb the climate for long periods of time.

---

We are going to discuss this in detail in the next chapter, but just quickly now, plants though photosynthesis take atmospheric $CO_2$ and use it to make their own energy and in the process create a waste stream of oxygen. It's a neat and self-sustaining cycle that forms the basis of life on Earth.

So what are the chemistry fundamentals here? The atomic

---

[67] *Here on Earth: an argument for hope.* Tim Flannery, 2010 Chapter 22. Text Publishing, Melbourne, Australia.

weights[68] of the three most abundant elements in the atmosphere are:

Carbon (C) is 12
Hydrogen (H) is 1
Oxygen (O) is 16

So $CO_2$ has an atomic weight of 12 + (2 x16) that is 44. The amount of carbon in a molecule of $CO_2$ is therefore 27 per cent. That is 12 is 27 per cent of 44.

Most common organic compounds in plants and animals have a structure based on variations of $C_1H_2O_1$. In the simplest form the atomic weight of common organic compounds is $C_1H_2O_1$ or 12 + 2 + 16 = 30. The amount of carbon in one molecule of a simple sugar therefore 40 per cent, that is, 12 is 40 per cent of 30. The simple sugar glucose has a molecular structure of $C_6H_{12}O_6$. The arithmetic is simple enough here.

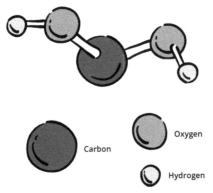

Most common organic compounds in plants and animals have a structure based on variations of $C_1H_2O_1$. In the simplest form the atomic weight of common organic compounds is $C_1H_2O_1$ or 12 + 2 + 16 = 30. The amount of carbon in one molecule of a simple sugar therefore 40 per cent.

---

68  The concept of atomic weight is fundamental to chemistry. The atomic weight of an atom is determined by the mass number of the atom, equal to the total number of protons and neutrons combined.

In many cases the oxygen component is slightly less than the simple model $C_6H_{12}O_6$. For example, $C_{12}H_{22}O_{11}$ is the formulation of sucrose and the amount of carbon is typically closer to 43 per cent and so on for other carbon-based organic substances. Some are really complex, such as xylon, a sugar contained in wood. We could get really technically complex here, but suffice to say that xylon can adopt several structures depending on conditions. Its chemical formula is: $HOCH_2(CH(OH))_3CHO$.

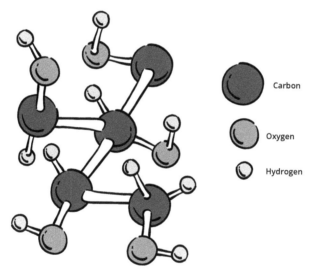

Some organic compounds, such as xylon, a sugar contained in wood are really complex. Xylon can adopt several structures depending on conditions. Its chemical formula is: $HOCH_2(CH(OH))_3CHO$.

---

Two chemical processes operate simultaneously in plants. Photosynthesis in the presence of energy from the Sun converts $CO_2$ and water into a simple sugar and oxygen is then released as a by-product. The chemical formula for photosynthetic reaction is:

$CO_2 + H_2O$ [+ sunlight] $= C_1H_2O_1 + O_2$

Respiration, or oxidation in chemistry terminology, is the second chemical process. It reverses the photosynthesis process and converts sugar and oxygen back to carbon dioxide and water. Energy is derived from the oxidation reaction.

Branches and leaves: The 'green' factory where photosynthesis converts $CO_2$ and water into sugars and oxygen.

Both plants and animals depend on photosynthesis to combine $CO_2$ and water to produce carbohydrates.[69] Breaking down carbohydrates through respiration is the mechanism used by both plants and animals to provide the energy required for growth and reproduction.

As previously indicated, carbon is present in our environment in a variety of different carbon sinks. It is dissolved in our oceans; in the biomass of plants or animals, whether living or dead; in the atmosphere, mostly as $CO_2$; in coal, oil reserves, limestone and other

---

[69] Carbohydrates are biological molecules consisting of carbon, hydrogen and oxygen atoms, usually with a hydrogen-oxygen atom ratio of 2:1. They are the sugars, starches and fibers found in fruits, grains, vegetables and milk-based products.

Photosynthesis and mineral cycles ... the process of photosynthesis results in the absorption of carbon dioxide, the production of oxygen and the storage of carbon.

rock types. Carbon is the absolute indispensable building block of life. You and I are made up of about 20 per cent carbon.[70] Plants have a much higher percentage.

———————————— • ————————————

Billions of years ago, when life on the planet was still embryonic, there was a lot more $CO_2$ in the atmosphere than there is today, for living things had not yet discovered a means to use it in the process of photosynthesis. Today, however, $CO_2$ forms just four parts per ten thousand of the gaseous composition of Earth's atmosphere, while the by-product of photosynthesis, oxygen forms 21 per cent. Flannery[71] describes the atmosphere's composition as:

... *the ultimate measure of life's triumph.*

---

70  Measured by dry weight – yes I know – then we would be dead.
71  *Here on Earth: an argument for hope.* Tim Flannery, 2010 Chapter 4, Page 45. Text Publishing, Melbourne, Australia.

Stored carbon reserves are largest in the tropical forest belt that circles the equatorial centre of the planet. Although tropical forests cover about five per cent of the Earth's surface, they are disproportionately high in importance in the planet's climate system. In addition to storing carbon they have a major influence on global weather patterns and also of course on biodiversity, with an estimated two-thirds of all living species residing in them.

Soils also represent a huge carbon reserve around 150 billion tonnes worldwide, which is roughly twice the amount of carbon in the atmosphere and three times as much as is contained in vegetation.[72] However, the world's intensively used crop lands have lost 30 to 75 per cent of their carbon content over the past two centuries.[73] That is around 78 billion tonnes of carbon. When combined with the carbon lost from poorly managed rangelands and from eroded soils (neither of which have been reliably estimated) it is clear that a huge amount of carbon has moved from soils into the atmosphere.

---

Each year the activities of human society contribute 7900 million tonnes of carbon to the atmosphere, of which the carbon sinks absorb 4600 million tonnes, leading to an annual net increase of 3300 million tonnes.[74]

This imbalance is so acute that it will not be enough simply to reduce carbon sources, carbon sinks will also have to be increased, and one of the easiest ways to increase carbon sinks is to increase the use of wood, and of course plant more trees.

There are just two ways to reduce $CO_2$ in the atmosphere; either by reducing emissions, or by removing $CO_2$ and storing it –

---

72 *Here on Earth: an argument for hope.* Tim Flannery, 2010 Chapter 22. Text Publishing, Melbourne, Australia.
73 *Here on Earth: an argument for hope.* Tim Flannery, 2010 Chapter 22. Pages 263-264 Text Publishing, Melbourne, Australia.
74 IPCC Assessment Report 2000 UN International Panel on Climate Change.

Although tropical forests cover only five per cent of the Earth's surface, they are disproportionately high in importance for the planet's climate system.

reducing *carbon sources* and increasing *carbon sinks*. Wood has the unique ability of being able to do both.

This chapter has been heavy on figures, so to prolong the anguish a bit more to finish. A simple example from a great little book: *The miracle of trees*[75] puts the process of photosynthesis into figures that provide a bit of real life context. The book explains that on a sunny day a 100 year old beech tree breathes in about 35,000 litres of air. From this air the tree extracts 10,000 litres or eighteen kilograms of $CO_2$ and produces twelve kilograms of sugars and thirteen kilograms of oxygen. The book asserts that walking in a forest on a sunny day you can literally feel the oxygen in the air.

We are going to explore the world of photosynthesis in more detail in the next chapter. Trust me, it's exciting!

---

75  From Page 12 of: *The Miracle of Trees*, Olavi Huikari 2012 Walker Publishing, New York.

CHAPTER 7

# Nature's magic show

How in a process called photosynthesis leaves turn carbon dioxide from the sky into giant trees

Apart from a very few organisms able to draw energy from hot ocean vents – all life on the planet depends either directly or indirectly on energy captured from the Sun. At the beginning of this chain are the plants, or more precisely their leaves that convert $CO_2$ and water into chemicals, like glucose that store the Sun's energy.

Trees have rightly been described as the biological miracle at the very core of human evolution. They provide the essential ingredients

Jack fruit leaves: At the most basic level, it is the leaves in the green canopies, where the truly magic story of nature is told.

that have given rise to the creation of life. At the most basic level, it is the leaves – in the green canopies – where the truly magic story of nature is to be found.

But wait a minute – if we are going to talk about trees – just what is a tree then? I think most would agree that the basic features of a tree are that they are typically tall and long lived. So trees are simply plants that have learnt to grow high and live for a long time. They grow tall to compete with other trees – racing upward and spreading outward for sunlight and water.

Also important, but less obvious, trees grow from the top. Some other plants that may also be tall, such as bananas and some palms

Canadian forest: A tree usually has a rigid, woody, strong expanding trunk, encased and protected by a layer of bark. These trunks support crowns of branches, twigs and leaves.

grow from their base – that is from the ground. They are not trees. A tree usually has a rigid, woody, strong expanding trunk, or trunks, encased and protected by a layer of bark. These trunks support crowns of branches, twigs and leaves.[76]

Trees also have a complex root system acting both to anchor the tree to the ground and to allow water and nutrients to be absorbed. With the constant imperative of seeking resources from the sky and the soil, given sufficient time, a tree can become huge – they just keep on growing.

Trees come in an amazing variety of forms from tall and narrow – as with many conifers – to the broad and spreading form of European oaks or African umbrella thorn trees. A tree's height and shape is determined both by its genetic 'blueprint' and its environment.

Trees are the largest organisms that have ever lived, with some giant specimens ten times heavier than a full-grown blue whale. Trees have dominated the land for more than 300 million years, far longer than the dinosaurs or mammals. They are far more diverse than animals, with many thousands of species living in a wide range of habitats.[77]

Trees are also the longest living organisms on the planet. Many live hundreds of years and some thousands of years. For example, the bristlecone pine, growing on the cold, dry mountains from the Mexican border north to Colorado in the United States of America, can live more than 4500 years. The world's longest lived trees include Australia's mountain ash, ancient huon pine and messmate.

In the big tree stakes New Zealand's kauri has to be included. Mature kauris are huge trees. Some of them are 2000 years, even 3000 years old. They are giants with tall cylindrical trucks topped by enormous spreading crowns. The kauri's exceptional height and

---

76  *Jungle Jive: Sustaining the Forests of Southeast Asia*, John Halkett 2016 Chapter 1. Connor Court Publishing, Redland Bay, Queensland, Australia.
77  *Talking Trees* blog. John Halkett. See: http://www.talkingtrees.com.au/oldest-and-biggest/

Bristlecone pines, Mount Evans, Colorado: Trees are also the longest living organisms on the planet. The bristlecone pine, growing on the cold, dry mountains from the Mexican border north to Colorado in the United States of America, can live more than 4500 years.

girth, together with its long, minimally tapering trunk combine to yield the greatest trunk volume of any tree. Yet they cannot claim to be the tallest, broadest or longest living tree in the world.

According to Fred Hageneder[78] in his 2005 book *The Living Wisdom of Trees*, the entire spectrum of human existence is reflected in tree lore through the ages; from birth, death and rebirth to the age old struggle between good and evil, and the quest for beauty, truth and enlightenment. He writes:

> *Whatever our personal beliefs regarding nature, spirits, and the question of whether God exists inside creation or only outside it (or at all), one thing is certain: the ability to extend compassion to other life forms, to feel gratitude and give thanks for sharing in the miracle of life, to respect, if not to love, all fellow inhabitants of this planet, makes us better human beings and helps us to triumph over ignorance and greed. The living wisdom of trees shows us that life is worth so much.*

---

78  Fred Hageneder is a recognised authority on botany and trees. He has written several tree books and has a tree-related website: www.themeaningoftrees.com

Hageneder suggests that the living wisdom of trees tells us that we are all travelling together through the cycle of life.

There is something about standing next to a really big tree, knowing that it is hundreds of years old – it somehow puts human life into perspective.

―――――――――――――  •  ―――――――――――――

Trees and humanity have always had a close working or symbiotic relationship. Animals consume leaves and breathe in the oxygen emitted by plants. In turn animals exhale carbon dioxide that plants require in order to carry out photosynthesis and form carbon-based sugars. This is a closed circuit that must be protected if life is going to continue.

The way trees grow is often not well understood. It is frequently thought that they grow from the ground, from their roots. However, this is not correct. Trees grow from the air through the process of photosynthesis – drawing in $CO_2$ through tiny stomata in their leaves. A neat way of summarising this process of photosynthesis that I like was written by Tim Flannery. He said[79]:

> We commonly misunderstand how trees grow, imagining that they somehow spring from the earth, from their roots. But this is not the case. Trees grow from the air, through tiny holes in their leaves call stomata, into which they draw $CO_2$.

Within the green, thin and vein-covered fabric of leaves something truly miraculous happens that provides the spark of life. It is the intrigue of how leaves in the process of photosynthesis are able to turn $CO_2$ from the sky into giant trees and other plants. How, within these green leaf factories the stuff we and other animals breathe out is combined with energy from the sun and water drawn up from the

―――――――――――――

[79] *Here on Earth: an argument for hope*. Tim Flannery, 2010 Chapter 22, Page 258. Text Publishing, Melbourne, Australia.

Microscope view of leaf underside stomata that open and close absorbing carbon dioxide, and releasing oxygen and water vapour.

ground and turned into carbon-based sugars and cellulose to make that renewable commodity wood. Yes it is truly nature's magic show that is the basis for life as we know it.

Photosynthesis literally means *light synthesis* – nature's way of capturing solar energy and storing carbon. Wood produced by photosynthesis and subsequent timber products store carbon throughout their lifetime – an important point we will return to in the pages ahead.

Photosynthesis takes place in specialised structures inside leaves called *chloroplasts*. Present in all green parts of plants, chloroplasts are where the work of capturing the sun's energy and converting it to sugars, takes place.

Looking a bit further at the process of photosynthesis – when rays of sunlight strike leaves, light wave lengths in the green spectrum bounce back – so we see green. The other light wave lengths – the reds, blues, indigoes and violets – are trapped in the chloroplasts where their energy is captured to break apart molecules of $CO_2$ in the production of simple sugars. These simple sugars are combined into larger, more complex cellulose or lignin sugars, or stored as energy reserves in starches and tubers. It is these complex sugars upon which the plants themselves, animals and indeed civilisations are built.

Without leaves capable of eating-the-sun we simply would not be here – plain and simple. Look at a tree and what you see is mostly congealed $CO_2$. The carbon atom in $CO_2$ is used by plants to build bark, wood and leaves – indeed all the tissues of the plants around us – a tonne of dry wood being the result of the destruction, by photosynthesis of around two tonnes of atmospheric $CO_2$. The two larger oxygen atoms are liberated into the atmosphere as a molecule of $O_2$.

David Suzuki explains the process of photosynthesis this way[80]:

*The sun is the primary source of energy for the enormous web of living things. Through photosynthesis, plants capture sunlight and convert photons of energy into stable molecules of sugar, where the energy is tied up in chemical bonds. In a miraculous bit of biological alchemy, plants inhale carbon dioxide from the atmosphere, add water from the soil, and, with energy from sunlight, create chains and rings of carbon that are the backbones of all the large carbon-based molecules of life.*

Green plants are much more efficient in their energy use than are fossil fuel power stations. Green plants manage to convert around one hundred billion tonnes of atmospheric carbon into living plant tissue each year, and in so doing remove eight per cent of all global atmospheric $CO_2$. This is an extraordinary figure, as if no additional $CO_2$ moved into the atmosphere, in just twelve years plants would have absorbed and used all of the atmospheric $CO_2$.[81]

---

[80] *The Legacy: An Elder's Vision for our Sustainably Future*, David Suzuki, 2013 Page 26 Allen & Unwin, Vancouver, BC, Canada.
[81] *Here on Earth: an argument for hope*. Tim Flannery, 2010 Chapter 4, Page 42. Text Publishing, Melbourne, Australia.

Nature's magic show  **71**

Tea plantation, India: Green plants convert around a hundred billion tonnes of atmospheric carbon into living plant tissue each year, and in so doing remove eight per cent of all global atmospheric carbon dioxide.

Photosynthesis stands at the heart of Earth's productivity. A leaf is a small miracle, for through it a lifeless gas is turned into a solid, living being.

Referring to David Suzuki again, he stresses the critical role photosynthesis plays in the removal of carbon $CO_2$ from and contribution of oxygen to the atmosphere. He says[82]:

> *So long as the carbon-based molecules created by photosynthesis are stored in the structures of trees and other plants, then that carbon remains bound up, or sequestered, and out of the atmosphere. Thus, the protection and preservation of photosynthetic activity must be an important consideration as we try to minimize the impact of climate change.*

---

82  *The Legacy: An Elder's Vision for our Sustainably Future*, David Suzuki, 2013 Pages 27-28 Allen & Unwin, Vancouver, BC, Canada.

Currently industrial methods of carbon capture are still under development and are yet to be operationally tested. There is no doubt that plants remain the most effective mechanism in existence for capturing carbon. Each year plants deal with 8 per cent of the atmospheric $CO_2$. Storing this captured carbon longer term is a trick we must learn. Trees and wood products are part of this conjuring act.[83] We will be spending time exploring these opportunities in some detail as we proceed through the pages ahead.

Another reality of global warming – curious but true – is that as the planet heats up, places that were once too cold for most plants to grow have become steadily more hospitable. This new vegetation exerts its own effects on the climate. According to a team led by Trevor Keenan of the Lawrence Berkeley National Laboratory, in California[84], the plant growth caused by climate change may also be helping to slow it – at least for now.

According to the Keenan-led team, between 1959 and 1989 the rate at which $CO_2$ levels were growing rose from 0.75 parts per million per year to 1.86 parts per million per year. Since 2002, this trajectory has stalled. In other words, although more $CO_2$ than ever is being pumped into the atmosphere, less than might be predicted of this excess is staying in the atmosphere. Keenan draws that analogy:

*Filling the atmosphere with $CO_2$ is a bit like filling a bath without a plug: the level will rise only if more water is coming out of the taps than is escaping down the drain.*

---

83   *Here on Earth: an argument for hope*. Tim Flannery, 2010 Chapter 22. Text Publishing, Melbourne, Australia.

84   *Recent pause in the growth rate of atmospheric $CO_2$ due to enhanced terrestrial carbon uptake* 2016 Trevor F Keenan, Colin Prentice, Josep G Canadell, Christopher A Williams, Han Wang, Michael Raupach and James Collatz Nature Communications

As we have already discussed, atmospheric carbon is removed and stored in carbon sinks, of which the oceans are one and photosynthesis is another, locking the carbon away in wood and leaves. Towards the end of the twentieth century around 50 per cent of the $CO_2$ emitted each year was removed from the atmosphere and stored in carbon sinks. Now, according to the Keenan-lead research, that number seems closer to 60 per cent. Carbon sinks seem to have become more effective, but the precise details are still unclear.

Using a mix of ground and atmospheric observations, satellite measurements and computer modeling, the Keenan team has

Green plants are much more efficient in their energy use than are fossil fuel power stations. Photosynthesis removes atmospheric carbon, locking it away in wood and leaves.

concluded that faster-growing land plants are the chief reason for this increased stored of carbon in sinks. The basic proposition is that as $CO_2$ concentrations rise, photosynthesis speeds up. Studies conducted in greenhouses have found that plants can photosynthesize up to 40 per cent faster when concentrations of $CO_2$ are between 475 and 600 parts per million.

However, Keenan's team sounds a warning – more vigorous photosynthesis is only slowing climate change and will not last. They state that there is more to growing plants than $CO_2$. They point to climate change making areas of high rainfall probably becoming wetter, while drier parts of the planet are becoming drier. Changing rainfall patterns could make some places less friendly to plants, and although plants benefit in the short term from extra $CO_2$, they suffer when temperatures get too high.

There will be other more complicated effects. For instance, much of the new plant growth has occurred in colder regions of the Earth and while ice and snow reflect sunlight, vegetation soaks it up, so more greenery in the north of the Northern Hemisphere will eventually lead to yet more warming. Elsewhere, higher temperatures could damage tropical forests. According to one estimate, for every degree of warming, tropical forests may release greenhouse gases equivalent to five years' worth of human emissions.

Summarizing the Keenan-lead research in the Energy and Environment section of The Washington Post Chris Mooney concluded[85]:

*Global greening, then, offers only a little breathing space. Kicking the fossil-fuel habit remains the only option.*

---

85 *If you're looking for good news about climate change, this is about the best there is right now* Chris Mooney
The Washington Post 10 November 2016

At this stage, if I may be so bold as to say so, I think some climate change scientists have been quite timid about expressing the present and prospective contribution tree plantations, natural forests, and wood products are capable of making to climate change abatement. I plan to be more positive in the chapters ahead and provide some evidence to support this position.

CHAPTER 8

# So what about forests then?

A key ingredient in balancing carbon accounts

Examining the use of forests, particularly intensively managed tree plantations, to both store carbon and provide an alternative energy source to fossil fuel is worthy of careful consideration.

The quantity of $CO_2$ that can be stored, or sequestered – to use a more impressive word that means the same thing – into biomass depends on the type of plantation or forest, growth rates, and the ultimate fate of the wood or other biomass sequestered and removed.

Giant redwoods, California, USA: A forest of mature trees 'locks up' large amounts of carbon, but little new growth takes place in such a forest.

A 'complication' to the tree-carbon storage issue is that it runs into a road block when trees approach maturity and their growth slows and eventually stops – trees do not grow on forever. A forest of mature trees 'locks up' large amounts of carbon, but little new growth takes place in such a forest. To store ever increasing amounts of carbon requires log harvesting, wood product manufacturing and tree regeneration or replanting. Even then the wood harvested needs to be stored long term – turned into durable timber products like houses and furniture.

It is an ecological reality that forests not used for wood production will transition from a period of growth when carbon is actively sequestered to a state when, as a mature ecosystem growth and decay balance out. As some trees begin to die and their crowns become less dense, the amount of decaying and dead matter increases and the forest may move from storing carbon to becoming a net producer of $CO_2$. Typically in mature forest ecosystems an equilibrium is reached when sequestered $CO_2$ and $CO_2$ production are in balance.

The essential point to make here is that natural forests tend move towards a state of equilibrium, where tree growth is balanced out by decay. So the net result is that no additional carbon is stored. This state of equilibrium may be disturbed, for example, by fire where the dramatic loss of $CO_2$ from burnt trees tends to be followed by a period of growth and net sequestration until a steady state of equilibrium returns once more.

Without getting too abstract and mystical, forests are one of the planet's great ecosystems. They are amazing communities of plant, animals, and ecological processes interacting with the environment. Like other types of ecosystems forests deliver life's critical ingredients – clean air, pure water, habitat and resources.

Forest cover varies enormously from one part of the planet to

Forest cover varies enormously from one part of the planet to another, it is largely climate that determines height, complexity, tree density and species composition.

another. Largely it is climate[86] that determines height, complexity, tree density and species composition. As forest or woodland types change so too does the whole assemblage of resident plants and animals.

To help round out the forests part of the story we should mention the very broad forest categories. Nature has gifted trees with the ability to survive in extreme weather conditions – they are found in all regions of the planet capable of sustaining plant growth. It is thought that forests cover about four billion hectares – about 30 per cent of the world's land surface.[87]

Simply forests can be described as occupying three broad zones, cold, warm and hot. These zones are essentially determined by

---

86 Notable among climatic variables that determine forest presence and characteristics are temperature, including both summer and winter averages and extremes, rainfall and its distribution over the seasons, humidity and wind.

87 *State of the World's Forests 2012*, Food and Agriculture Organization of the United Nations, Rome 2012.

So what about forests then? **79**

Although commonly based on latitude, forests can also be classified by dominant tree species, resulting in different forest types.

temperature with a close correlation with latitude. These zones are replicated on both sides of the equator with the exception of cold zone forests. Although forests were once present on the Antarctic continent[88] it is now the exclusive domain of ice and penguins.

Forests can also be classified by other characteristics – the options are wide and potentially confusing. Although commonly based on latitude, forests can also be classified based on the dominant tree species, resulting in numerous different forest types, such as ponderosa pine, beech forest and so on.

In the north of the Northern Hemisphere we have the austere

---

88  Discussion about the evidence of former forests in Antarctica is outlined in Chapter 8 of: *Southern beeches*, A.L. Poole, 1987, Science Information Publishing Centre, Wellington, New Zealand.

yet spectacular boreal[89] forests. These forests occupy the northern sub arctic zone up beyond about latitude 50 and are generally comprised of evergreen conifer tree species. The boreal region encircles the earth at the top of the Northern Hemisphere across Russia, Scandinavia, Alaska and Canada. The boreal forest belt represents the world's largest land based ecosystem – a swath of confers, with some deciduous[90] trees that act as part of the largest source and filter of freshwater on the planet. Beyond the northern limit of boreal forest lies bleak treeless arctic tundra and ice.

Pine forest: As forest or woodland types change, so to does the whole assemblage of resident plants and animals.

Temperate forests occupy the zone between the chilly northern boreal forests and the tropical forests of the equatorial zone. The

---

89  The word 'boreal' means northern. Boreal forests are also known by the Russian name *taiga*.
90  Deciduous trees shed their leaves at the end of a growing season and grow them again at the beginning of the next growing season. Most deciduous trees have flowers and broad rather than needle-like leaves and are found predominantly in temperate forests.

climate across temperate zone forests is neither extremely hot nor really cold. The summers can feel hot and dry, but in these forests climate is never so harsh that soil dries up or plants die. Likewise, the winters may produce a lot of snow, but not as severe as in boreal forest regions.

Temperate forests are found in both hemispheres from latitudes 25 to 50 in regions of north eastern Asia, North America, western and central Europe, southern South America and Australasia. They can be coniferous, deciduous or contain mixed species depending on geography and climate.

Temperate forests include some of the world's tallest trees, such as the giant redwoods along the northwest coast of North America.

Northern temperate deciduous forest: Temperate forests are found in both hemispheres from latitudes 25 to 50 in regions of north eastern Asia, North America, western and central Europe, southern South America and Australasia.

Other major areas of temperate conifer[91] forest include the Northern Hemisphere forests of the mountains of western China, northeastern China and adjacent regions of Russia. Also in Japan, the mountains of central Asia and in the Himalayas, temperate conifer forests are present. Mexico's Sierra Madre ranges, central Europe the Balkans and Turkey all possess conifer dominated forests as do the more mountainous regions of the Atlas ranges of northwest Africa.

In the Southern Hemisphere there are smaller regions of conifer forests, such as the monkey puzzle tree, or *Araucaria* and the tall, long lived *Fitzroya cupressoides* conifer native to the Andes mountains of southern Chile and Argentina; the kauri and podocarp forests of New Zealand and some small patches of ancient conifers in Tasmania, Australia.

Vast tracts of former temperate deciduous dominated forest have been cleared for settlement and farming. Frequently derived from glacial deposits the underlying young soils are mostly fertile and moisture retentive. In large parts of China this clearance has been extensive; less so in Europe. In the United States of America settlers have cleared vast tracts of deciduous forest for cities and farms over the past four centuries.

In the Southern Hemisphere, broadleaf[92] temperate forest includes the denser forests in cooler regions of South America – really only in Chile and Argentina, in Australia, New Zealand and the southern tip of Africa. In Australia eucalypts account for more than 70 per cent of trees in forests and woodlands, growing in a wide range of climates from the hot tropics to near desert inland plains and up on to alpine snow fields.

---

91  Conifer means 'cone bearer' and refers to the way conifer trees carry their seeds on the scales of cones, like pine, fir and spruce trees, rather than enclosed in a fruit developed from a flower.
92  Broadleaved trees are flowering plants. That is they have 'normal' leaves as opposed to conifers, like the pines that have narrow needle-like leaves.

In Australia eucalypts account for more than 70 per cent of trees in forests and woodlands, growing in a wide range of climates from the hot tropics to near desert inland plains, and up on to alpine snow fields.

Tropical forest forms a discontinuous band around the Earth, bisected somewhat unequally by the Equator, so that rather more tropical forest lies in the Northern than the Southern Hemisphere. A particular feature of tropical forest is that the overwhelming majority of plants are trees of all shapes and sizes. Not only are trees the dominant members of these communities, but most of the climbing plants and some of the epiphytes are also woody. The undergrowth largely consists of seedling, sapling trees, and young woody climbers.

The trees of the tropical forest are numerous in terms of the number of species present. They are sometimes taller than those in temperate forests, but they do not reach the gigantic dimensions of Californian redwoods or the large eucalypts of Australia.

By way of further illustrating the contrast between forest types, in temperate forests dominant trees frequently belong to just a few and sometimes a single species. By comparison in tropical rainforests there is seldom less than around 50 trees species present on any given hectare and frequently well over a 100. The richness and diversity of the tree flora is indeed an important characteristic of tropical forests.

Now returning to the discussion we were having about forests storing carbon, or not depending on the stage of the growth cycle. Where biomass is removed, or harvested from time to time forests continue to sequester carbon to replace the carbon contained in the biomass removed. When a forest is harvested there will be a period while

Confer forest on the slopes of Skrzyczne in southern Poland: It is estimated that the forests across the European Union, together with the related wood products sector, currently contribute to climate change mitigation to an amount equal to 13 per cent of total Europe Union carbon emissions.

the remaining and regenerating, or replanted trees grow to occupy the space vacated by the extracted trees. If a forest is completely removed and subsequently regenerated or replanted, there will be a period while trees are regrowing when carbon sequestration will be low or even negative. The essential point here is that only by continuing to grow a forest or plantation and remove biomass from time to time in the form of wood can the forest continue to indefinitely sequester carbon.

As an example, it is estimated the forests across the European Union, together with the related wood products sector currently contribute to climate change mitigation at an amount equal to 13 percent of total European Union country carbon emissions.[93]

Softwood tree plantation: The amount of carbon that can be stored in a plantation depends on the rate of growth and the species involved. Higher rates of carbon storage are achieved by well managed, fast growing species on good soils with adequate rainfall areas.

---
93  *European State Forests Boost the Bioeconomy* 2016 EUSTAFOR, Brussels.

This includes the forests' carbon sink function, as well as carbon stored in wood-based products, and the use of biomass 'waste' to substitute for fossil fuels.

The amount of carbon that can be stored in a forest or plantation depends on the rate of growth and the species involved. Higher rates of sequestration are achieved by well managed, fast growing species on good soils with adequate rainfall areas. Lower rates of carbon sequestration occur where rainfall is low, soils are not optimal, management is poor and slow growing species are planted.

However, the importance of forests stretches far beyond their boundaries, because forests help regulate the planet's climate. For example, according to the United Nations Food and Agricultural Organization they store nearly 300 billion tonnes of carbon in their living parts – roughly 40 times the annual greenhouse gas emissions from fossil fuels.[94]

As we have already acknowledged, there are large regions of the world where the march of civilization – settlements, gardens, fields and animals – has over the past couple of thousand years cleared large tracts of forest.

By 2009, around half of the tropical forests present around 1800 had been cleared[95] and indeed are continuing to be cleared. As a result some developing countries have high rates of greenhouse gas emissions. Papua New Guinea, for example, produces a third as much greenhouse gas as Australia, a country with four times the population and that burns substantial quantities of coal to generate electricity.

---

94  *The Role of Old-Growth Forests in Carbon Sequestration* 26 July 2016 Geoffrey Cragg, Future Directions International Pty Ltd. Western Australia. See: www.futuredirections.org.au
95  *Here on Earth: an argument for hope.* Tim Flannery, 2010 Chapter 22, Pages 260. Text Publishing, Melbourne, Australia.

The most recent consolidated data for global forest cover shows a net loss of 129 million hectares of forest between 1990 and 2015, resulting in a one per cent reduction in forest land as a proportion of the global land area. However, the rate of annual net loss of forest has slowed from 0.18 per cent in the 1990s to 0.08 per cent in the period 2010–2015.[96]

Globally, forests provide wood, food and income to billions of people. Over three billion cubic metres of wood are harvested from forests annually. About 2.4 billion people cook with wood fuel, and at least 1.3 billion people rely on forest products for shelter. Forests also support vibrant industries, formally employing about 13.2 million people across the world and informally employing at least another 41 million.[97]

Globally 15 per cent of all human-caused greenhouse gas emissions result from the destruction of tropical forests. If we could reverse this tragic trend, and by 2050 restore at least between 8 and 17 per cent of what we have removed, then between 40 billion and 200 billion tonnes of $CO_2$ could be sequestered in this rehabilitated forest.[98]

---

96 *Global Forest Resources Assessment 2015: How are the world's forests changing?* FAO. 2016a 2nd ed. Rome.
97 *State of the World's Forests 2014: Enhancing the socioeconomic benefits from forests.* FAO. 2014 Rome.
98 This would make an impressive contribution towards balancing the carbon account book.

CHAPTER 9
# Selling the carbon to save the trees

Emissions trading – fighting climate change and protecting forests

Remember back in Chapter 2 we discussed the adverse and depressing contribution forest destruction – particularly of tropical forests – has made to increasing global $CO_2$ emissions. Just to recap, over the last 200 years, the concentration of $CO_2$ in the atmosphere has jumped by 30 per cent and this extra carbon has come from two sources. Around 40 per cent from the destruction of forests and associated soils, and the rest from the burning of fossilised carbon in the form of coal, oil and gas.[99]

In relation to forest destruction, between 200 and 250 billion tonnes of carbon has been emitted into the atmosphere over about the last three centuries making up around 30 per cent of all the carbon released into the atmosphere over that period.[100]

In pursuit of arresting climate change it is obvious that we need to do all we can to keep trees standing, and of particular importance, tropical forests intact. Also it is the easiest option for confronting climate change – so really a no-brainer.

●

As we have already outlined, in an endeavour to tackle climate change $CO_2$ emissions trading schemes have been developed and

99  *How Coal Works* Union of Concerned Scientists. www.ucsusa.org/clean_energy
100  *Here on Earth: an argument for hope.* Tim Flannery, 2010 Chapter 15, Pages 197-198. Text Publishing, Melbourne, Australia.

Jungle clearing is a principle reason for biodiversity loss. Every effort needs to be made to keep trees standing, and importantly tropical forests intact.

implemented in some countries. An emerging option in relation to such trading schemes is the ability to commercially trade – read sell – carbon stored in forest or plantation trees. The fundamental requisite of such trade is that the forest remains standing. Quantifying, valuing and trading carbon stored in trees would, in addition to providing a source of income, protect forests. So positives all around, with substantial collateral benefits for environmental values and animal inhabit retention.

An example of an emissions trading scheme is the *Reduced Emissions through Deforestation and Degradation* (REDD) scheme. This scheme provides a means of paying to keep tropical forests standing by essentially selling the carbon stored in trees – in trunks, even roots, and in some cases the soil – as carbon *credits* or *offsets*. The revenue regenerated from the sale of such carbon credits or

Amazon rainforest, Tambopata Reserve, Peru: There must be strong economic incentives to retain and perpetuate tropical forests, rather than destroy them. Such economic incentives will help to create a set of circumstances where tropical forests are seen as an economic asset, not a liability.

offsets can be spread across the supply chain – if you like – from governments to companies with legal entitlements over the land, right down to local communities and traditional indigenous owners. By doing so we can change the dynamic from a view that says tropical forests are economically 'worthless' to one that sees their retention as commercially attractive.

It is apparent that there must be strong economic incentives to retain and perpetuate tropical forests, rather than destroy them. Such economic incentives will help to create a set of circumstances where tropical forests are seen as an economic asset, and not a liability.

Emissions trading schemes are sometimes referred to as cap-and-trade schemes. Typically such schemes involve a government entity setting a limit, or *cap*, on the amount of $CO_2$ that can be emitted. The limit or cap can be allocated or sold to companies in the form of emissions permits that essentially represent the right to emit – or discharge – specific volumes of $CO_2$. Under emissions trading schemes these offsets assist industrial $CO_2$ emitters required to progressively reduce their emission levels in accordance with a national scheme mandating that emissions levels be kept under a prescribed 'cap'.

Emitters of $CO_2$ that increase the volume of their permitted emissions, or intend to keep them under a reducing cap, are required to purchase carbon offsets, or to pay a 'fine' to the government. The transfer of carbon offsets is referred to as a *trade*. In effect, the buyer is paying a charge for polluting, while the seller is being rewarded for generating carbon offsets, from say preventing forests being cleared or by planting trees. It is really a smart cycle that both addresses net carbon emissions and keeps trees and forests standing.

The development of effective carbon credit or offset trading mechanisms that allows $CO_2$ emitters to enter into commercial contracts with forest growers needs some fundamental elements. These include the need to be able to quantify the amount of carbon stored in a defined area of forest or tree plantation, and legally robust certification to facilitate the issue carbon credits that permits them to be traded and audited, providing adequate assurance to purchasers and government regulators.

To a significant degree the opportunity for forests and tree plantations to have a positive role in climate change abatement depends on making progress in arresting deforestation and forest degradation. As mentioned earlier, the REDD scheme envisages payments to forest owners and other parties for trading carbon

To a significant degree the opportunity for forests and tree plantations to have a positive role in climate change abatement depends on making progress in arresting deforestation and forest degradation.

stored in trees and for conserving forests. However, while REDD schemes seem to offer enticing prospects[101] there is still remaining uncertainty related to how such schemes will be further developed and implemented, and to what extent they will become an important component of climate change mitigation and forest protection. Obviously stepping up tree planting to enhance carbon storage is another sensible and valuable strategy.

The key conclusion of significant research is that the relative differences in the greenhouse gas balance of 'production' and 'conservation' scenarios for forest management is that policies halting forest management for wood production are not warranted.

---

101 For further information go the Intergovernmental Panel on Climate Change website: http://www.ipcc.ch/

Eucalypts plantation: Stepping up tree planting to enhance carbon storage is a sensible and valuable strategy.

When industry value-added and carbon abatement benefits are added together, the production management scenarios for forests generate much higher greenhouse gas mitigation values than do conservation management scenarios alone.

There is considerable room, however, for improvement in the greenhouse gas outcomes of managing forests for production. These opportunities could be realised in the processing of wood products, and in diverting materials to increased use of biomass for energy production, which will be discussed as you read on. Increased use of forest and sawmill residues for renewable energy generation typically displaces the use of fossil fuels, resulting in a net greenhouse gas benefits.[102]

This all reaffirms that in an increasingly carbon constrained

---
102 *Forest HWP – A new approach to the assessment of the carbon cycle in native forests managed for multiple use.* February 2016 Also see www.fwpa.com.au/rdworks-newsletters/936-foresthwp-a-new-approach-to-the-assessment-of-the-carbon-cycle-in-native-forests-managed-for-multiple-use

world, forests and wood products are important not only as carbon sinks and to generate carbon offsets, but as substitutes for more carbon intensive materials and fossil fuels. Sustainable forest management and the increased uptake of wood products should play a greater role in efforts to reduce greenhouse gas emissions.

As we will discuss in the chapters ahead, wood products can also contribute to 'avoided' greenhouse gas emissions because only a relatively small amount of energy is used in their production, compared with other building materials such as brick, concrete, steel and aluminium.

CHAPTER 10

# Getting the good wood message

Employing the power of photosynthesis

I think it will be helpful at this point in proceedings if we pause for a moment before launching into the chapters ahead. So far I have attempted to set the scene, discussing, as we have, the climate change challenge. This has included some background commentary; what is climate change; the greenhouse effect; the basics of the underlying chemistry, and nature's miracle of photosynthesis. So far I have attempted to capture a sense of the dire issues around the science-

Eucalyptus trees employing the wonder of photosynthesis to do things that are practical and achievable to address the climate change threat.

verified climate change trends, and likely future climate change-related impacts. We have also briefly skated over trees and forests, saying something about their characteristics and distribution.

In the chapters ahead we will focus on how we can employ the wonder of photosynthesis to do some things that are practical and achievable to address the climate change threat we have identified and discussed. This is not to say that other measures are not also important – they are. However, others have covered this ground much more competently than I am capable of doing. Plus some solutions are technologically very complex feats of engineering, so I will stick to the things I know a little about – trees and wood – they are more my interest, and much less complex measures.

We discussed the nature and attributes of trees – nature's

Basically wood is really just congealed $CO_2$. As well as giving trees support their wooden trunks contain tissue through which water and nutrients are distributed and stored.

architectural and engineering marvels in the last chapter – so enough about them for now. However, because I intend to talk about the ability of wood products to store carbon and contribute towards tackling climate change I should say something about wood itself before proceeding further.

Basically wood is really just congealed $CO_2$, although that is a bit of an over simplification. As well as giving trees support as they grow taller, their wooden trunks contain tissue through which water and nutrients are distributed and stored.

Bark is the outermost layer of tree trunks. Underneath this outer, protective bark layer is the inner bark, cambium, sapwood and the central heartwood.

The cambium layer is the area of the tree trunk where new cells are formed progressively adding to the amount of sapwood in the tree trunk and expanding the girth of the trunk. As the area of sapwood expands, the older sapwood toward the centre of

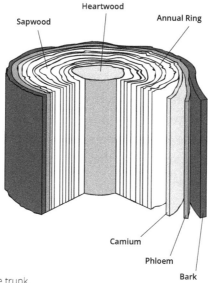

Cross section of a tree trunk.

the trunk becomes less active, darker in colour and referred to as heartwood.

The sapwood region of the trunk is the powerhouse of the tree. It is through this actively growing sapwood tissue that moisture and minerals are carried. The cells in the sapwood involved in transportation up and down the trunk are the *xylem* and *phloem*. Xylem cells conduct moisture and dissolved minerals from the roots upwards and outwards throughout the tree. The cell walls of the xylem are strengthened with a fibrous material called lignin.

Phloem cells in the sapwood are responsible for the downward transportation of carbon-based sugars manufactured in the leaves during the process of photosynthesis that we discussed earlier. These sugars are transported the full length of the tree from its leaves to its roots. Cells known as rays, found in the phloem, also carry sugars inward further into the sapwood where they are converted to starch and stored. If the outer layers of a tree trunk, particularly the cambium and sapwood are damaged, for example by ring barking or fire, the tree will suffer and may die.

It is the darker and harder, once xylem and phloem sapwood towards the centre of trunks, that is termed heartwood. As sapwood cells break down with age and use and become heartwood the tree forms new sapwood. The role of heartwood is to give the tree strength and stability. The darker coloration in heartwood, compared to the outer sapwood, is due to its impregnation of the former sapwood cells with oils, gums and resins that protect the heartwood from insect attack and fungal decay.

---

As I will attempt to illustrate, wood is an important weapon in the fight against climate change. In addition, it offers some new and exciting opportunities to extend its use as a building material

Getting the good wood message   99

Wood provided the material used for the construction of primitive dwellings and stockades by early human society and the stuff from which weapons were fashioned for hunting and defence.

beyond traditional uses and family homes.

As we know the use of wood by human society is not new. Wood was a source of fuel for our distant ancestors as they learnt to use fire and to cook. Wood provided the material used for the construction of primitive dwellings and stockades and the stuff from which weapons were fashioned for hunting and defence. Wood was also important in the early industrial forging and manufacturing of metals like bronze, copper and iron.

———————————    •    ———————

In the pages ahead we will have a look at both existing and potentially new applications of wood in home and building construction. We will also look at the current and the future potential to use wood as a source of energy to replace fossil fuels to assist in heading us in the direction of a future zero net carbon world.

I also want to say something about the opportunities for carbon credit trading. This mechanism has the potential to assist in reversing disappointing trends in tropical forest destruction. As outlined in Chapter 4 the concentration of $CO_2$ in the atmosphere has increased by 30 per cent, and about 40 per cent of this increase is a result of the destruction of forests and soils. So retention and expansion of tropical forests would make a very substantial contribution towards reducing $CO_2$ emissions.

As you read on we will discuss the benefits to climate change abatement of a collaborative effort between trees and wood. Then finally I will attempt to outline some practicable suggestions as to how we can use the power of photosynthesis through trees, forests and wood to fight the threat that climate change presents to the survival of us all.

CHAPTER 11

# Working together

Global agreements set the pace

To paraphrase a saying – the road to environmental good intentions is paved with agreements. Perhaps not surprisingly as the 'big' environmental challenges facing the planet do not recognise national boundaries – wild fires and atmospheric pollution in Indonesia impact on Singapore and Malaysia, and fossil fuel induced elevated atmospheric $CO_2$ levels impact on the security of small island nations as sea levels rise. So international consensus and agreement is needed to attempt to come to terms with global issues, including climate change.

Going back in time, environmental issues first registered on the international community radar because of the compelling

Rachel Carson: Her book *Silent Spring* exposed the environmental consequences of the indiscriminate use of pesticides – notably the insidious effects of DDT.

advocacy of people like Rachel Carson. In her book *Silent Spring*[103] she exposed the environmental consequences of the indiscriminate use of pesticides – notably the insidious effects of DDT[104] and more generally also important questions about humanity's impact on nature. The book described how DDT entered the food chain and accumulated in the fatty tissues of animals, including human beings, causing cancer and genetic damage. Rachel Carson's seminal book concluded that DDT and other pesticides had irrevocably harmed animals and had contaminated the world's food supply and accused the chemical industry of spreading misleading information and public officials of accepting false industry claims.

The book spurred a reversal in United States pesticide policy, led to a nationwide ban on DDT for agricultural uses, and inspired an environmental movement that led to the creation of the US Environmental Protection Agency.

•

The issue of the relationship between economic development and environmental degradation was placed prominently on the international agenda by the end of the sixties decade. An early key global gathering was the United Nations Conference on the Human Environment held in Stockholm, Sweden in 1972. By 1983, mounting concerns about global environmental 'stress' resulted in the Brundtland Commission[105] making the term 'sustainable development' part of modern vocabulary. Essentially this marriage between previously strange bedfellows – economic and environmental outcomes – is now central element of acceptable natural resource management.

---

103 *Silent Spring* Rachel Carson 1962 Houghton Mifflin, New York, USA.
104 DDT is an abbreviation of dichlorodiphenyltrichloroethane. It is a colorless, crystalline, tasteless and almost odorless known for its insecticidal properties and environmental impacts.
105 Formally known as the World Commission on Environment and Development (WCED), the Brundtland Commission's mission was to attempt to unite countries to pursue sustainable development objectives.

Now with the recognition that deforestation and tropical forest degradation are resulting in significant environmental damage, the role of forests in climate change mitigation is becoming much more centre stage in global climate change discussion and politics.

The tug-of-war over the science and policies associated with $CO_2$ emissions and doing something constructive about tackling climate change was initially centred around two interrelated agreements; the 1992 United Nations Framework Convention on Climate Change and the 1997 Kyoto Protocol.

Both initiatives attempt to set out a framework for scientific and political co-operation. The central idea of the United Nations Framework Convention is that there is a common duty to: *prevent dangerous human-induced interference with the climate system*, essentially the unrestricted pumping of $CO_2$, and other greenhouse gases into the atmosphere. The Kyoto Protocol goes considerably further, than its predecessor and established legally binding minimum emissions reduction obligations for industrial countries.

•

In the Japanese City of Kyoto, governments agreed to what was subsequently called the Kyoto Protocol that commits industrialised countries to reduce or limit their greenhouse gas emissions and reach certain emission reduction targets by 2012.

Recognizing that developed countries are principally responsible for the current high levels of greenhouse gas emissions as a result of more than 150 years of industrial activity, the Kyoto Protocol placed a heavier burden on developed nations under the principle of common but differentiated responsibilities. The Kyoto Protocol focused on industrialised countries because they are responsible for most of the past and current greenhouse gas emissions and have the knowledge and resources to reduce them.

In the Japanese City of Kyoto, governments agreed to what was subsequently called the Kyoto Protocol that committed industrialised countries to limit their greenhouse gas emissions and achieve certain emission reduction targets by 2012.

The Kyoto Protocol came into force in 2005 and was formally adopted by 183 governments. The protocol sets emissions targets for 37 industrialised countries. Most of these targets require emissions reductions of 5-8 per cent from 1990 levels by 2012. Among industrialised nations, only the United States of America decided not to participate.

The protocol also introduced various economic mechanisms under which countries could co-operate in reducing emissions. For example, the *Clean Development Mechanism* allows industrialised countries to meet their emission targets partly by investing in emission-saving projects in developing countries. This in turn is helping to transfer new emission reduction technologies to poorer countries. The Kyoto Protocol was the first global environmental investment and credit scheme of its kind. A parallel mechanism, known as *Joint Implementation*, enables industrialised countries to invest in such projects on each other's territory.

The Kyoto Protocol is constrained because developing countries have so far 'escaped' without any emission limitations. This situation was considered untenable long term as emissions from 'emerging' countries will exceed the capacity of the atmosphere to absorb them, even if industrial country were to completely cease emissions.

Fundamental to the framing of the Kyoto Protocol was a recognition that the climate change debate throws up a number of questions such as how much of the pace of climate change can be attributed to human influences and what degree of global warming is acceptable? Answers to these and similar questions are critical in determining upper emission limits, and acknowledging amongst other things, that the damaging effects of climate change will not hit everyone equally. For example, rising temperatures

Because they are responsible for most of the past greenhouse gas emissions, and have the knowledge and resources to reduce emissions, the Kyoto Protocol focused on industrialised countries.

seriously threaten the economic security and the culture of Inuit people of Canada's arctic region. Hunting lifestyles are disappearing as traditional routes across the ice become unusable. Inuit food supplies are threatened, permafrost breaks up and igloos lose their insulating properties.

Also if you are living on a low-lying Pacific island like Tuvalu or parts of the Solomon Islands group where the highest point is just a few metres above sea level you will most certainly start to worry about your future prospects.

---

Negotiations in relation to a revised set of commitments under the Kyoto Protocol took place in Paris in December 2015. This widely published gathering of world leaders and thousands of supporting experts and advocates culminated in 187 countries agreeing to limit global warming to 1.5 degrees Celsius. This agreement, thought unthinkable just a few months earlier, is a substantial advance on the two degrees Celsius target that around 200 countries had agreed as a limit six years ago in Copenhagen.[106] But let's wait and see.

At the commencement of the Paris talks countries submitted pledges to cut or curb their carbon emissions. However, collectively these pledges were not sufficient to prevent global temperatures from rising beyond two degrees. In fact several analyses suggested they may still see global temperatures rise to three degrees.

The Paris meeting resulted in countries promising to attempt to reduce global emissions down from peak levels as soon as possible.

---

106 The Copenhagen Climate Change Conference took place in Copenhagen and was hosted by the Government of Denmark. Close to 115 world leaders attended the high-level segment, making it one of the largest gatherings of world leaders ever outside Unite Nations meetings. In addition more than 40,000 people, representing governments, nongovernmental organisations, intergovernmental organisations, media and United Nations agencies attended. The conference produced the Copenhagen Accord that contained several key elements on which there was strong convergence of the views. This included the long-term goal of limiting the maximum global average temperature increase to no more than 2 degrees Celsius above pre-industrial levels, subject to a review in 2015. There was, however, no agreement on how to do this in practical terms.

More significantly, they pledged[107]:

> ... to achieve a balance between anthropogenic emissions by sources and removals by sinks of greenhouse gases in the second half of this century.

In more understandable language this means getting to "net zero emissions" between 2050 and 2100. The United Nation's climate science panel argued that net zero emissions must happen by 2070 to avoid dangerous global warming.[108]

The Paris Agreement recognises the key role of forests in meeting this challenge, not only for their mitigation potential, but also for their contribution to adaptation.

In order to meet the intention of the Paris Agreement, countries are called to contribute to emission reduction within their different capabilities, on the basis of voluntary commitments expressed through Intended Nationally Determined Contributions (INDCs). The Paris Agreement recognised that forests and wood products offer both developed and developing countries a wide range of options for timely and cost-effective mitigation.

While the Kyoto Protocol defined legally binding responsibilities for only industrialised countries, the Paris Agreement calls on all countries to contribute to the mitigation effort, respecting their different capabilities and responsibilities. Once a country signs the Paris Agreement, the INDCs become Nationally Determined Contributions.

More than 70 per cent of the countries that submitted INDCs include forests as an important component of their contribution to climate change mitigation.

In relation to forests text of the climate 'pact' released at the conclusion of the Paris meeting, encouraged countries to take action

---

107 Paris COP21: *Key issues for the new climate agreement*. Climate Council 2016 www.climatecouncil.org.au
108 Base on media reports including those in The New York Times. See: www.nytimes.com/interactive/2015/12/12/world/paris-climate-change-deal-explainer.html?_r=0

Redwood forest: More than 70 per cent of the countries that submitted 'Intended Nationally Determined Contributions' include forests as an important component of their contribution to climate change mitigation.

to implement and support positive incentives directed at reducing emissions from deforestation and forest degradation, and support the role of conservation, sustainable management of forests, and enhancement of forest carbon stocks.[109] Emphasis was placed on policy approaches that supported integrated, sustainable management of forests, while at the same time recognising the importance of actions and incentives to safeguard non carbon benefits associated with such approaches to forest management and protection.

A valuable outcome of the Kyoto Protocol was that the protocol fostered a carbon market. With accounting rules and project guidelines for the generation of carbon credits, it defined the activities eligible for mitigation and helped to shape investments in climate change mitigation in developed and developing countries.

---

109 United Nations, *Tackling Climate Change*. See:www.un.org/sustainabledevelopment/climate-change/

Globally, however, the combined value of carbon pricing instruments was less than US$50 billion in 2015, of which almost 70 per cent was attributed to emission trading systems and the rest to carbon taxes. Carbon prices vary significantly, from less than US$1 to US$130 per tonne of carbon. About 85 per cent of emissions are priced at less than US$10 per tonne. This is considerably lower than the price estimated as needed to meet the recommended two degrees Celsius climate stabilisation goal.

Almost 20 years after the Kyoto Protocol was agreed, the United Nations Framework Convention on Climate Change framework has only recently begun to provide more opportunities for unlocking forest mitigation potential. Both developed and developing countries can now contribute to mitigation, and financial support can now be provided not only through transaction of emission reductions, but also through other instruments of climate and development finance.

———————————— • ————————————

To a large degree the opportunity for forestry to have a positive role in climate change abatement and the provision of environmental services depends on making progress in arresting deforestation and forest degradation. The REDD scheme first mentioned in Chapter 9 envisages payments to forest owners and other parties for trading carbon stored in trees and for conserving forests. However, while REDD schemes seem to offer enticing prospects[110] uncertainty remains as to how such schemes will be developed and implemented, and to what extent they will become an important component of climate change mitigation and forest protection. Stepping up tree planting to enhance carbon storage is obviously another valuable strategy.

———————————— • ————————————

110 For further information go the Intergovernmental Panel on Climate Change website: http://www.ipcc.ch/

Soldier and school kids planting trees, Thailand. Stepping up tree planting to enhance carbon storage is another valuable strategy.

Noteworthy in the Paris Agreement is the recognition of the importance of forests for both climate change mitigation and adaptation. The finance section of the agreement singles out REDD activities in particular, calling on countries to support REDD and alternative policy approaches such as joint mitigation and adaptation approaches for the integral and sustainable management of forests. The Paris Agreement also affirms the importance of non-carbon benefits, which could pave the way for reflection of the multiple benefits from forests in future carbon prices.

In their statements at the signature ceremony of the Paris Agreement more than 40 developing countries highlighted their mitigation actions in the forest sector, and in REDD in particular. Of 185 INDCs submitted, 122 countries, or 76 per cent mentioned forests. Forests were mentioned in INDCs for all regions, but most frequently in Africa, where forests were mentioned in mitigation by 92 per cent of countries and in adaptation by 75 per cent of countries. REDD opportunities contributing to both adaptation and mitigation, were mentioned specifically by 28 countries. Many more list actions that would fall under REDD, such as avoiding emissions from deforestation and land degradation through sustainable forest management or the enhancement of carbon sinks. Forestry is mentioned in 82 per cent of INDCs submitted by developing countries and in all of those submitted by the least developed countries, indicating the important role of forests in bringing developing countries to the global Paris Agreement.

The majority of developing countries make their INDCs contingent on financial and technical support from developed countries. Some countries call for immediate increased contributions

Pine plantation and log dump, Sunny Corner State Forest, New South Wales, Australia. Forestry is mentioned in 82 per cent of INDCs submitted by developing countries, and in 100 per cent of those submitted by the least developed countries under the Paris Agreement.

to the Green Climate Fund and stated that international funding sources should include reliable, new and additional development assistance, and not redirected development funds.

———————— • ————————

It might sound a bit boring, but global agreements and coordinated global action is absolutely necessary to come to grips with actions needed to confront climate change. There is no doubt however that binding international treaties have in the past provided difficult to negotiate, and perhaps even more difficult to implement and enforce. But let us be clear, in the case of climate change, humanity has little option other than to endeavour to work together on actions agreed in Paris – our collective survival depends on it.

CHAPTER 12
# Feeling better now?
Health and wellbeing benefits if you have wood in your life

While the primary focus of this book is on the contribution trees, forests and wood are capable of making in tackling climate change, it is also worth pointing out some of the other remarkable qualities of nature's miracle 'inventions' – trees and wood.

As if being renewable, storing carbon and contributing to climate change mitigation isn't enough – there's more – live in a wooden house and you will feel better – true! It is not stretching the point to say your health and well-being are likely to be improved if you live in a wooden environment. A study by Planet Ark[111] has lent weight to arguments that wooden buildings create a healthier living environment.

The study *Wood – Housing, Health, Humanity*[112] reviews worldwide studies and feedback from citizens. The study concludes that, across housing, business, education, and healthcare buildings, "multiple physiological, psychological and environmental benefits are identified for wooden interiors". People also appeared "innately drawn towards wood, which elicits feelings of warmth, comfort and relaxation, and creates a link to nature".

Given today's trend towards urbanisation and high-density apartment living the Planet Ark study is particularly relevant. People are now less exposed to open spaces and nature that have demonstrated benefits to cognitive abilities, self-esteem levels, lower stress response, blood pressure and cholesterol. Wood-based building can induce similar physiological and mental responses.

---
111 Established in 1992, Planet Ark promotes "environmental behaviour change" to help the human population "live in balance with nature".
112 *Wood – Housing, Health, Humanity* 2015 Planet Ark, Australia. See the report at: http://makeitwood.org/benefits-of-wood/wood-for-health.cfm

Wooden interior: People are innately drawn towards wood, which creates a link to nature, elicits feelings of warmth, comfort and relaxation.

The Planet Ark study cites Japanese research that demonstrates exposure to wood panelled interiors has a blood pressure lowering effect, while steel raises it. Reports from Austria showed sleeping in a bed from Stone pine[113] reduced subjects' heart rate by 3500 beats a day, while wooden classroom interiors reduced perceptions of stress in high school students.

Planet Ark says findings were particularly significant for children with high stress levels at a young age linked to mental disorders later in life.

•

Further Japanese analysis indicates that wood could also be good at the other end of the age spectrum. Timber interiors resulted in elderly health centre residents "willing to engage and interact more

---

113 Stone pine (*Pinus pinea*) also called Italian stone pine, umbrella pine and parasol pine is native to the Mediterranean region, occurring in Southern Europe, Israel, Lebanon and Syria. The tree has been cultivated throughout the Mediterranean region for so long it has naturalized, and is now considered native beyond its natural range. It is also present in North Africa, the Canary Islands, South Africa and Australia. It is also planted in Western Europe up to Scotland, and on the East Coast of the United States of America up to New Jersey.

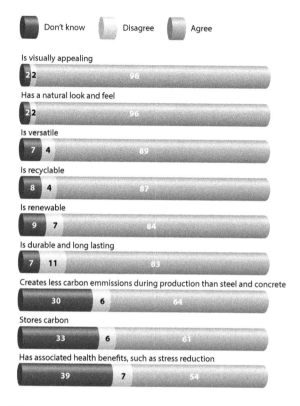

Results of the Planet Ark survey on whether Australians 'agree', 'disagree' or 'don't know' when asked questions about wood (Source: Plant Ark).

with one another", while their "positive emotional state expanded self-expression in positive way".[114]

In Australia, 96 per cent of respondents to a Planet Ark questionnaire found wood interiors "visually attractive". Wood products within a room have also been shown to improve indoor air quality by moderating humidity.

From a science perspective, the body of knowledge about these health and well-being benefits is growing to the point that a number

---

114 *Influence of wood wall panels on physiological and psychological responses.* Sakuragawa, S., Miyazaki, Y., Kaneko, T. and Makita, 2005 51, 136–140.

Wooden buildings, Huilo, Patagonia, Chile: Wood increases feelings of happiness and elevates self-esteem levels.

of architects and designers are specifically designing schools and health care facilities with appreciable amounts of exposed wood. The award winning Dandenong Mental Health Facility[115] is a case in point. The designers of the facility specifically chose wood, both new and recycled, to provide warmth, texture, patterning, tactility and a non-institutional feel.

The Planet Ark study found that wood elicits feelings of warmth, comfort and relaxation and creates a link to nature. Further, that wood increases happiness and self-esteem levels; increases cognitive abilities, and decreases stress response, blood pressure, pulse rates and cholesterol levels. These benefits are particularly important for

---

115 The Dandenong Mental Health complex is the largest mental health facility in Victoria. It consolidates all residential, research, training and administration functions into a purpose-built precinct facility.

The extensive use of timber internally in walls and ceilings and externally in cladding and exposed structure to provides warmth, texture, patterning, tactility and a non-institutional feel. It also gives the buildings a residential, suburban character which is enhanced by carefully landscaped courtyards and external garden areas avoiding the need to construct high institutional walls

The building was the winner of the Australian Institute of Architects National Architects Award 2014 and the 2014 Australian Timber Design Award.

environments where it is difficult to incorporate nature indoors, like hospitals where strict health and safety guidelines may prevent the presence of plants, and office environments where views from the window are of neighbouring concrete buildings.

So it turns out that timber buildings aren't just healthier for the environment but for the humans who live in them. This is further reinforced by a 2009 study by Austria's Human Research Institute[116] showing that children taught in timber buildings experience less stress and concentrate better. Further, that timber-built retirement villages promote tranquillity and other benefits.

Exterior of a wooden terraced apartment building. Timber buildings are not just healthier for the environment, but for the humans who live in them.

---

116 Joanneum Research Institute of Non-Invasive Diagnosis.

Jonathan Evans[117] says there is now a push in the United Kingdom and Germany for homes for the elderly to be built out of wood. "There is a strong belief that timber buildings provide a more serene environment; they feel more human," he says.

Promoting these benefits to the community, home owners, designers and architects, is therefore of significant value. Increasing urbanisation means that people have less access to nature in their daily lives. Australians on average now spend about 90 per cent of their time indoors.[118] This coincides with reports of increasing obesity and nearly half of Australians experience a mental health condition during their lifetime.[119] As it is not always possible to increase time spent outside, particularly in workplaces, schools and hospitals, understanding how to incorporate the physiological and psychological benefits of nature into indoor environments is an increasingly important area of research.

Wood products within a room have been shown to improve indoor air quality by moderating humidity. This effect occurs because wood absorbs moisture from the air in humid conditions and releases moisture in dry conditions.[120]

Because research has demonstrated that productivity is reduced by an average of 12 per cent in offices where staff are dissatisfied with the quality of the air, the ability of wood to moderate humidity is has a particularly beneficial impact in workplaces.[121]

The Planet Ark study concludes that when compared to other material types the positive views of wood continue, as outlined in

---

117 Architect quoted in *Sydney Morning Herald, Good Weekend Magazine* article: *New Wood: how it will change our skyline*, Greg Callaghan 27 August 2016.
118 *Indoor air.* at <http://www.environment.gov.au/topics/environment-protection/ air-quality/indoor-air>
119 *Global, regional, and national prevalence of overweight and obesity in children and adults during 1980-2013* Ng, M. Globan Burden of Disease Study 2013. 384, 766–781 (2014).
Depression and anxiety are common conditions. See; http://www.beyondblue. org.au/the-facts>
120 Effect of healthy workplaces on well-being and productivity of office workers. Bergs, J 2002.
121 *Global, regional, and national prevalence of overweight and obesity in children and adults during 1980-2013* Ng, M. Globan Burden of Disease Study 2013. 384, 766–781 (2014).

Blackwood strung chairs: Wood products in a room improve indoor air quality by moderating humidity.

the diagram on the next page. Wood was viewed as the material that creates a natural look and feel, warm and cosy environments, is visually appealing, and is nice to touch, by nine out of ten people.[122] It is also viewed as being the most environmentally-friendly material by seven out of ten people. By comparison the second most popular material, brick, received 34 per cent less positive feedback. Plastic was seen as the cheapest material, but it scored lowest in four out of five categories related to creating pleasant surroundings and being environmentally-friendly.

———————————— • ————————————

Even though many people don't understand the health and well-being benefits of wood, they instinctively react to the feelings of warmth and comfort it creates and its natural look and feel. An increasing body of research is concluding that being surrounded

---

122 *Effects of relaxing music on state anxiety in myocardial infarction patients*. Bolwerk, C. A. L 1990 13, 63–72. *Therapeutic influences of plants in hospital rooms on surgical recovery*. Park, S.-H. & Mattson, R. H. 2009 44, 102–105.

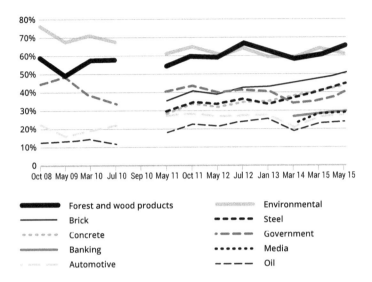

| | | | |
|---|---|---|---|
| ▬▬▬ | Forest and wood products | ░░░░ | Environmental |
| ——— | Brick | ▬ ▬ ▬ | Steel |
| ∙ ∙ ∙ ∙ | Concrete | ∙ ▬ ▬ ▬ | Government |
| ▬▬▬▬ | Banking | ∙ ∙ ∙ ∙ ∙ ∙ | Media |
| — — | Automotive | – – – – | Oil |

Evidence of an increasing awareness of the carbon storage properties of wood and wood products showing a progressive trend increase in consumer understandings for the topics of carbon storage in wood, the permanence of carbon storage in wood and the role of wood products in the home to store carbon (Source: Pollinate (2015) Project Toy Story Key Findings).

by wood at home, work or school has positive effects on the body, sense of well-being and the environment. So increasingly architects who design buildings for healing, learning and relaxation are incorporating wood into their structures to capitalise on its health and well-being benefits.

The positive attributes towards wood from the Planet Ark study are reinforced by Forest and Wood Products Australia Limited (FWPA) commissioned research[123] undertaken to evaluate their generic wood marketing activities carried out to promote the use of wood and wood products.

The research found that preferences and emotional attachment to wood are, and have remained high since 2007-2008. Where an

---

[123] *Forest and Wood Products Australia Generic Marketing promotions* – Evaluation, October 2015, Centre for International Economics, Canberra, Australia.

understanding of the environmental properties of wood and wood products was not as universally held in the early years of the FWPA marketing campaign, increased awareness is considered to have had a distinct effect on indirect demand, as well as on measures of social licence.[124]

Consumer tracking studies have also reported increased trust attributed to the forest and wood products industry. This is likely to have a strong positive effect on the social licence for the sector. In general, when compared to other materials, such as steel, concrete and bricks, wood is found to lead the way in terms of nearly all positive emotional and attitudinal characteristics.[125]

There is evidence of an increasing awareness of the carbon storage properties of wood and wood products. The diagram on the next page shows a progressive trend increase in consumer understanding of the topics of carbon storage in wood, the permanence of carbon storage in wood, and the role of wood products in the home to store carbon.

Other research findings from tracking studies include that almost 60 per cent of consumers (up from a low of 45 per cent in 2010) agree that more wood products should be used because wood is 'more environmentally-friendly' than other building products. Also there is an increasing international trend of consumer willingness to pay for the environmental performance of products.

Market investigations have found that a lack of information about wood is a key impediment to it being specified more in building and design projects. Increased understanding of wood and wood products is known to be a driver of increased specification, with

---

[124] A social licence is considered to exist when a project or activity has the ongoing support of the local community and other stakeholders, ongoing approval and/or broad social acceptance.

[125] *Forest and Wood Products Australia Generic Marketing promotions – Evaluation*, October 2015 Centre for International Economics. See; http://www.fwpa.com.au/about-us/corporate-documents/882-forest-and-wood-products-australia-generic-marketing-promotions.html

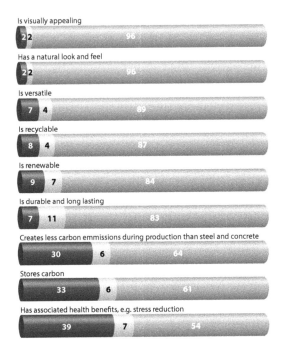

Evidence of an increasing awareness of the carbon storage properties of wood and wood products showing a progressive trend increase in consumer understandings for the topics of carbon storage in wood, the permanence of carbon storage in wood and the role of wood products in the home to store carbon (Source: Pollinate (2015) Project Toy Story Key Findings).

a survey of American architects noting that increased knowledge leads to a higher potential for specification. An increased awareness and understanding of wood and wood products amongst building professionals leads to the conclusion that there was likely to have been a market effect due to the FWPA *WoodSolutions* campaign.[126]

So there is some solid research reinforcing the not entirely unexpected connection between well-being and nature – in this

---

126 WoodSolutions website : www.woodsolutions.com.au

Wooden beams: More wood products should be used because wood is 'more environmentally-friendly' than other building products. There is also an increasing trend of consumer willingness to pay for the product environmental performance.

case specifically wood. Just think about it – you can love a beautiful piece of wooden furniture or polished wooden floor, but it is hard to have the same feelings about a concrete beam or a plastic chair.

CHAPTER 13

# Is the sky the limit?

Inwards and upwards, drivers behind city
building innovation.

When we think of climate change, we mostly think of the dangers – like higher temperatures, rising sea levels, bush fires and more severe storms. We don't think of a potential economic upside, but that's what the building and construction sector has within its reach.

It's exciting times for tall wooden buildings in today's cities. In many cities population growth, space availability and infrastructure pressures are all contributing to the need for innovative building solutions. It seems that inwards and upwards, plus the need for strong environmental credentials, are key drivers behind future city building activity. An aspect of this direction, that in part takes advantage of timber's dexterity as a building material and environmental credentials, is the development of a new generation of super strong engineered timber products.

Leading United Kingdom architect Alex de Rijke[127] argues that if the nineteenth century was the century of steel, and the twentieth century the century of concrete, then the twenty first century will be about engineered timber. Australian construction industry advisor David Chandler adds that while timber is a foundation construction material, it has not generally been considered as a viable alternative to concrete, steel and masonry. "This may be about to change. Engineered wood products could be at the cutting edge of that change," he said.[128]

---

127 Alex de Rijke is a founding director of the United Kingdom architectural practice dRMM. His work is well-known for innovative construction technologies and materials. He has taught at the Architectural Association, London; the Aalto University, Helsinki and as Guest Professor at the School of Architecture in Düsseldorf. The Royal College of Art has appointed Alex de Rijke as Dean of the School of Architecture.

128 Wood and Australia's 2023 construction road map, 2013 David Chandler. See: www.constructionedge.com.au

New generation, super strong engineered timber products. Taking advantage of timber's dexterity as a building material and its environmental credentials.

Besides its popular appeal, and lightness making it suitable for construction in densely built city centres, the range of other environmental benefits for building in wood are substantial. For a start there are significant $CO_2$ savings to be made by using timber in the construction of houses and other buildings, both in terms of embodied energy and in-use energy efficiency, and associated greenhouse gas emissions.

So just as steel, glass and concrete revolutionised super-tall building construction in the twentieth century it is now likely that timber –vastly kinder to the environment, faster to build, and with next-to-zero waste – will do the same as the new century marches on.

Fifth generation Australian timber man Chris Taylor[129] is at the leading-edge of innovative, expanded timber product use. In

---

129 Chris Taylor, is the part owner and managing director of Blacktown Timber in Sydney, Australia (see: http://bttimber.com.au/

relation to the expanding use of wood for both traditional and high rise building and construction he said:

*For contemporary architecture timber is the material of the moment, offering design opportunities well beyond the reach of alternatives. Covering modern timber architecture with stunning images of structural designs that vary from bridges to dwellings, walkways over water or a gazebo near the ocean, nothing equals the beauty, strength and durability of timber.*

---•---

It is apparent that Australia's largest cities are in the midst of a revolution. New apartment construction now outstrips traditional detached home building. Housing is certainly moving in towards city centres and development hubs, and up into the sky, and it is apartments not houses, that are important dwellings behind today's city building boom.

Australia's largest cities are in the midst of a revolution. New apartment construction now outstrips traditional detached home building. Housing is certainly moving in towards city centres and development hubs.

Another emerging feature of city apartment building in Australia is an increasing number of developments showcasing wooden mid-rise[130] buildings. Until recently the choice of building materials for mid-rise construction was limited to traditional 'heavy' construction materials generally excluding timber, but changes are now apparent. Whether its timber's natural aesthetic attributes, engineering properties, durability or its carbon sequestering credentials, wood is gaining traction as the building material for city renewal and expansion.

The trend to tall timber building is spreading globally. Last year the TREET tower in Bergen, Norway, designed by Artec was completed, reaching a new timber building world record with its 14-storeys. Also in 2015 three new towers were given the go ahead in the USA after winning the government's *Tall Timber* competition.[131]

*Haut*, a 73 metre skyscraper to be built in Amsterdam's Amstelkwartier, is another example of the growing timber architecture trend hitting tall building design. The project was awarded to a group of architects including ARUP, Lingotto, Nicole Maarsen, TEAM V Architecture and Nederlandse Energie Maatschappij. Construction began in late 2017.

The 21-storey building will house approximately 55 apartments. According to TEAM V, *Haut* stands for 'haute couture', tailor-made architecture: "the design offers buyers extensive freedom of choice in the size of their apartment, the number of floors, the lay-out and the positioning of double height spaces, and outdoor balconies. The wooden finishing underneath the balconies and the pronounced projections at the sharp building corner facing the River Amstel make *Haut's* architecture strikingly distinctive".[132]

---

130 In the context of this discussion mid-rise is considered to be apartments, hotels and commercial buildings up to eight stories, or to 25 metres in height.

131 *Announcing the U.S. Tall Wood Building Prize Competition to Innovate Building Construction* 10 October 2014, Doug McKalip, Senior Policy Advisor for Rural Affairs, White House Domestic Policy Council. See: www.whitehouse.gov/blog/2014/10/10/announcing-us-tall-wood-building-prize-competition.

132 *Tallest wooden building in the Netherlands to be built in Amsterdam* 25 July 2016. See: www.arup.com/news/2016_07_july/25_july_tallest_wooden_building_in_the_netherlands_to_be_built_in_amsterdam

ARUP says: "building in wood is one of the most talked about innovations in sustainable construction internationally, due to the large storage capacity of $CO_2$. Using wood provides an answer to the Municipality of Amsterdam's quest for $CO_2$ neutrality". The building is to receive the BREEAM[133] Outstanding label, the highest possible sustainability score.

According to ARUP *Haut's* wood can store over three million kilograms of $CO_2$. "In addition 1.250 square metres of solar panels

Building in wood is one of the most talked about innovations in sustainable construction due to the large storage capacity of $CO_2$ and sustainability score.

---

133 BREEAM is a leading sustainability assessment method for master planning projects, infrastructure and buildings. It addresses a number of lifecycle stages, such as new construction, refurbishment and in-use. BREEAM inspires developers and creators to excel, innovate and make effective use of resources. The focus on sustainable value and efficiency makes BREEAM certified developments attractive property investments and generates sustainable environments that enhance the well-being of the people who live and work in them.

will help the building produce renewable energy, while waste water is purified through a constructed wetland on the roof. The parking garage in the building has space for shareable electric cars".

Tall wooden buildings are, of course, nothing new. Japan's five-storey Horyuji Buddhist temple in Nara, built more than 1400 years ago, has survived earthquakes, fires and floods. But in most countries, timber lost its allure – especially for the construction of tall buildings once the art of stonemasonry, and later bricklaying and precast concrete took hold.

Tall wooden buildings are, of course, nothing new. Japan's five-storey Horyuji Buddhist temple built more than 1400 years ago, has survived earthquakes, fires and floods.

The potential of timber in city construction has prompted the unveiling of *woodscraper* proposals in Australia and elsewhere. Renowned for championing the tall wooden building industry, Canadian architect Michael Green suggests wooden skylines could reach 30 storeys and that we have the technology in place to build them right now.[134]

"Unlike steel and concrete, wood sequesters carbon dioxide, storing it away for the life of the building. As a renewable material grown by the power of the sun, wood offers us a new way to think about our future," he said.

Green confirms building such ambitious structures with timber will mean: "reinventing wood; making it stronger, more fire safe, more durable and selecting material from sustainably managed forests."

Structural engineer Nick Hewson agrees, but offers a caveat. "I think timber could definitely have a role to play in 30 storey plus

Wood-based building under construction in Europe: There is probably a wooden building 'sweet spot' between four and 15 storeys.

---

134 View Michael Green's TED lecture at: https://www.ted.com/talks/michael_green_why_we_should_build_wooden_skyscrapers?language=en

buildings, but they are unlikely to be entirely timber structures. Certain issues can arise when you start to build over ten storeys."[135]

"I think there's probably a wooden building 'sweet spot' between four and 15 storeys for a wholly timber building – the range where it will be most effective," he said.

According to Hewson, wooden buildings can weigh 50 per cent less than traditional concrete buildings. "This can extensively reduce the cost, particularly in cases where developers are working with poor ground conditions, or where a build site needs extensive outlay on foundations. Particularly in dense cities, where conditions are restrained and space is limited, timber becomes attractive. It is lightweight and easier to handle."

Hewson forecasts that the next five years will be filled with opportunities to apply wood as decking over railway lines. He identifies Australia's Melbourne as a city with an array of potential development sites around train stations. "There are such high costs in deck structure so if you can double the yield by placing a building on top, the economics start to make real sense."

"Timber can help manage a site better. Sites are quieter, less dirty, less dusty and significantly safer. They're also quick to construct with a timber building able to be up in a matter of weeks," said Hewson.

In Sydney, Australia, Urban Growth NSW[136] has identified an urban renewal opportunity for the Central to Eveleigh railway corridor which would see over a million square metres of floor space made available along the three kilometre corridor providing real potential for timber construction.

---

135 *Code change creates exciting new opportunities according to architects and engineers* Code change – architects' and engineers' comments 5 February 2016. See: www.woodsolutions.com.au
136 Urban Growth NSW is the NSW Government's urban transformation agency oversighting a range of projects supporting the government's strategic priorities, notably in the central business district and Western Sydney.

An industry-sponsored symposium held in Melbourne in 2016 discussed the rapidly growing opportunities in the mid-rise apartment, hotel and commercial buildings. The symposium was attended by representatives of major development companies, architects and building professionals.

The managing director of FWPA Ric Sinclair noted that the objective of the symposium was to provide an opportunity for the timber industry supply chain and other interested parties to explore possible market development approaches to the rapidly expanding mid-rise market for timber-based building solutions.

"The symposium followed a submission for *Proposal for Change* for the *National Construction Code* to make it easier to build mid-rise buildings up to eight storeys out of lightweight and massive timber construction systems".

Now easier to build mid-rise buildings out of lightweight massive timber construction systems, providing opportunities for structural and appearance, sawn, engineered and panels; massive timber and lightweight structural timber products.

These new markets provide opportunities for all timber products, both local and imported; structural and appearance, sawn, engineered and panels; mass timber and lightweight structural timber products," said Sinclair.[137]

Canadian timber industry expert Kelly McCloskey told symposium attendees that British Columbia got the ball rolling on building code changes for five and six storey wood buildings in 2009 and more than 250 projects are now built or near completion; "North America was perhaps 20 years ahead on multiple storey timber buildings, but Australia had the ability to catch up."

At the symposium IndustryEdge's Tim Woods said projected levels of higher density housing represents a major transition in

Cross laminated timber construction, Melbourne, Australia. The mid-rise market opportunity will be characterised by a greater evolution to 'systems-based' solutions and supply arrangements – both lightweight timber members and plasterboard, massive timber and prefabrication.

---

137 Major new opportunities with mid-rise timber buildings – application, specification, products and systems Mid-rise timber building seminar May 2016 See: www.woodsolutions.com.au

the way Australians live in cities, with potentially about a million new higher density homes being added to the housing stock in the next three decades through urban renewal and infill in Sydney, Melbourne and Brisbane alone.

"The mid-rise market opportunity will be characterised by a greater evolution to 'systems-based' solutions and supply arrangements – both lightweight timber members and plasterboard, mass timber and prefabricated products."

"A number of major builders have already tested a range of new timber systems in this market and found the approach to be highly cost competitive against current alternative building options," said Woods.

There seems little doubt that engineered timber products and building systems now available, along with clever design and changes to building codes will see wooden buildings continuing to move skywards. Smart marketing of the environmental and carbon storage attributers of wood, will add to this momentum.

Changes to the Australian National Construction Code from 1 May 2016 enable timber buildings of up to eight storeys, or 25 metres. Ric Sinclair said these changes would deliver a wide range of benefits to local residents, property buyers and the domestic building industry.[138]

> This initiative will bring Australia up to pace with much of the rest of the world – so that the building property industry can take advantage of the environmental and cost benefits of domestic timber construction.
> Wood can offer quicker build times, with less noise and disruption for neighbours. It can also offer innovative design approaches.
> A look at international trends shows the global sector is embracing

---
138 *Timber set to rise to eight storeys (and cut costs) under national construction code changes in May*. New code changes exciting news for design and construction 30 January 2016. See: www.woodsolutions.com.au

*both traditional wood and modern engineered wood products in an increasingly broad range of structural and decorative applications.*

Architects, developers, designers, engineers and builders have welcomed changes to the National Construction Code that apply to both traditional timber framing and innovative massive timber systems – such as cross laminated timber[139] and glulam.[140]

James Fitzpatrick of Sydney architects Fitzpatrick and Partners said: "This is an exciting step forward for architects and their clients. It not only gives us new material options to create innovative design solutions for our clients, but it also enables us to deliver

Changes to the National Construction Code provide new timber-based material options to create innovative design solutions delivering more environmentally advantaged and sustainable developments of apartment, office and hotel buildings.

---

139 Cross laminated timber or CLT comprises multiple layers of wood glued together at right angles under high pressure to form large wall, ceiling and floor panels.

140 Glued laminated timber, or glulam, is a type of structural engineered wood product comprising a number of layers of timber bonded together. By laminating a number of smaller pieces of timber together, a single large, strong, structural component is manufactured from smaller pieces. Glulam is used as vertical columns or horizontal beams, as well as curved, arched shapes. Glulam is readily produced in curved shapes and it is available in a range of species and appearance characteristics to meet varied end-use requirements.

more environmentally advantaged and sustainable developments. Ultimately, the code change will potentially mean quicker, more cost effective and environmentally-friendlier construction of apartment, office and hotel buildings."[141]

•

Economic modeling suggests these code changes will result in potential savings in the order of up to 15-25 per cent depending on build type, primarily due to shorter construction times. The modeling also suggests net benefits to the Australian economy over ten years of approximately $103 million; comprising $98 million in direct construction cost savings, $4 million in reduced compliance costs; and a million in environmental benefits.

Andrew Waugh of Waugh Thistleton Architects says Australian cities, like others round the world, need to embrace the opportunity. "The growth in the world's population is principally concentrated in urban areas."[142]

"Changes to the National Construction Code will allow architects to better explore and demonstrate the potential of engineered timber, while also increasing the use of traditional timber framing. This will allow the pioneering work done by Lend Lease on their Forté building to be emulated throughout Australia," said Waugh.

He predicts Australian architects will create more residential buildings where the timber and joinery are left bare and allowed to be the 'heroes' of the design. The structure might be exposed offering "an honesty" in the construction for all to enjoy.

"In Australia, as elsewhere, we need to increase the density of our cities. We also need to reduce our reliance on heavy, polluting

---

141 Code change creates exciting new opportunities according to architects and engineers Code change – architects' and engineers' comments 5 February 2016. See: www.woodsolutions.com.au

142 *Code change creates exciting new opportunities according to architects and engineers* Code change – architects' and engineers' comments 5 February 2016. See: www.woodsolutions.com.au

## Is the sky the limit? 137

Cross laminated timber panels. Energy efficiency and ecologically sensitive design will be 'drivers' of innovation in building materials and construction methodologies.

construction materials. Engineered timber allows us to build tall residential buildings in urban contexts, without costing the earth!" he said.

Nick Hewson is also a fan of the building process saying that construction in timber is typically significantly faster and quieter than more traditional construction methods.

This will provide opportunities to build using timber around sensitive locations, reducing disruption to surrounding facilities – perhaps next to hospitals or schools. "The more sensitive and constrained the site the more economic timber construction will tend to be."

In terms of urban design of middle suburbia, he said the lightweight nature of timber will be a huge benefit when designing developments over rail lines and stations. "There are huge amounts

Lend Lease's trail-blazing ten storeys Forté. When built it was the world's tallest high-rise apartment building using mass timber construction.

of valuable real estate across our major cities above existing rail lines and in Australia, we are lagging behind the rest of the world in capturing the value in these spaces."

Australia's most successful and high-profile residential building using engineered timber products is Lend Lease's trail-blazing *Forté*. At ten storeys, it was initially the world's tallest high-rise apartment building using massive timber construction. Based in Victoria's inner-city Docklands, its 'pre-cast' cross laminated timber (CLT) panels allowed fast, light and cost-effective construction.

Based on his observations in Europe, principal of Victorian architectural company Maddison Architects, Peter Maddison predicts there will be fewer buildings wrapped in aluminum or cement sheeting, corrugated iron or brickwork.

"There are plenty of precedents out there and Australia has been a bit slow off the blocks but the situation is changing rapidly and we're watching a natural evolution. It's such a viable way of building and is so sustainable."[143]

Hewson is particularly passionate about the potential for schools to be designed innovatively.

"As Australian suburbs start to become ever denser and more facilities need to be built, we'll see a change in the traditional school construction to more multi-storey buildings or 'vertical' schools," he said. Prefabricated timber construction will allow entire buildings to go up during school holidays. As well, he is excited by research that suggests children are happier, learn more and are relaxed surrounded by natural materials.[144]

Both Hewson and Maddison agree there are few impediments to medium-rise construction in timber other than those based largely on preconceptions and prejudices.

Hewson said he hoped this change will start to win over the hearts and minds of more people in the construction industry." My own experience has shown that once people have experienced a large-scale timber construction project, they are often converted and wonder why they didn't build that way before."

He said he was very passionate about reducing humanity's impact on the planet and there is a huge role for timber to play in that. "A project we're working on in south-west Sydney is just about to start on-site and we have calculated that the amount of $CO_2$ saved by using a timber structure compared to a concrete one is equivalent

---

143 Code change creates exciting new opportunities according to architects and engineers Code change – architects' and engineers' comments 5 February 2016. See: www.woodsolutions.com.au

144 *Code change creates exciting new opportunities according to architects and engineers* Code change – architects' and engineers' comments 5 February 2016. See: www.woodsolutions.com.au

Docklands Library and Community Centre, Melbourne, Australia: CLT and reclaimed hardwood cladding construction. Once people have experienced a large-scale timber construction project, they wonder why building hadn't been constructed that way before.

to removing 500 cars from our roads for a year. If we can build more buildings using sustainably-sourced timber then we can start to try and offset some of the damage we're doing to the planet by other means".[145]

The first commercial office building in Australia made entirely from timber was built by Lend Lease as the gateway to its multi-billion dollar Barangaroo development in Sydney. Completed towards the end of 2017 International House Sydney is a six-storey building made from engineered wood – both CLT and glulam – setting a new Australian benchmark in the use of sustainable building materials. The building was designed by Sydney architect Alec Tzannes,

[145] *Code change creates exciting new opportunities according to architects and engineers* Code change – architects' and engineers' comments 5 February 2016. See: www.woodsolutions.com.au

Is the sky the limit? **141**

International House Sydney the gateway to the multi-billion dollar Barangaroo development is the first commercial office building in Australia made from timber.

of Tzannes Associates. He said he aimed to create a new form of beauty beyond shape and surface. "It is 'deep design', renewing architecture's role to serve the greater social purpose of lowering carbon emissions," he said.

A striking colonnade of hardwood columns front the street. Above that, all the floors are CLT, the cores such as lifts are also CLT, while the columns and beams are glulam and exposed around the lifts. Raking columns are made of ironbark timber reclaimed from old bridges.

A report by the Australian Sustainable Built Environment Council[146] sets out detailed economic modelling showing how reducing emissions in the building sector will save money, grow jobs and improve quality of life.

Buildings produce almost a quarter of Australia's $CO_2$ emissions, but unlike the transport or aviation sectors, the technology exists to massively reduce building emissions. This provides the building and construction industry with a real and positive opportunity to contribute to climate change abatement.

Prefabrication of building components also has the potential to reduce construction time and costs, while enabling off-site construction of building elements and permitting high levels of customisation – improving building quality and value.

The Australian Sustainable Built Environment Council report says Australia's building sector can deliver up to 28 per cent of Australia's 2030 emissions reduction target, save $20 billion and create healthier, more productive cities if a suite of targeted policies are introduced.

The benefits are not just economic and environmental. Research shows that buildings designed for low emissions are also just better buildings. Features like more natural light and more comfortable temperature make low carbon buildings cheaper to light, heat and cool, more productive, healthier places to live, work and learn. We will explore these benefits further in the next chapter.

The Australian Sustainable Built Environment Council report sets out the key steps that policy makers need to take to overcome these barriers and achieve the necessary transition, including that governments should establish a national plan towards zero carbon buildings by 2050 that should include strong mandatory minimum standards for buildings, equipment and appliances. They could also use

---

146 *Low Carbon, High Performance. How buildings can make a major contribution to Australia's emissions and productivity goals*, May 2116. Australian Sustainable Built Environment Council www.asbec.asn.au

their considerable power as purchasers and provide financial incentives for building owners to implement emissions reducing initiatives. Council president Professor Ken Maher said:

*The good news is that major improvements are possible with the right public policies. Every year we delay will cost us significantly in emissions, climate change, money, and quality of life. Installing inefficient equipment or appliances locks in excessive emissions for many decades into the future. Even five years of delay in the take-up of these opportunities could lead to $24 billion in wasted energy costs and more than 170 megatonnes of lost emissions reduction opportunities*

*Key findings of the report shows that buildings account for 23 per cent of Australia's emissions, so strong action in buildings is essential to meet our international obligations to transition to zero net emissions by around 2050. Buildings can achieve zero carbon by 2050 using existing technologies. In addition to $20 billion in energy savings, buildings can deliver one quarter of the national emissions target and over half of the national energy productivity target by 2030.*

Eco-friendly green 'living' roof. Buildings can achieve zero carbon by 2050 using existing technologies.

Property Council of Australia chief executive Ken Morrison said the report was a blueprint for government action. "Major emissions reduction gains can be made, but it requires a focused plan that includes regulation, strong incentives, energy market reform and market information to support transformation."

Chief executive of the Green Building Council of Australia Romilly Madew said Australia consistently tops international tables for green building leadership. "We have more than 1000 low-carbon, Green Star-rated buildings around the country. While buildings generate 23 per cent of Australia's carbon emissions, we have the technology, the skills and the knowledge to halve emissions, while also boosting the productivity, health and wellbeing of the people who live, learn, work and play in our buildings."

———————————— • ————————————

Excuse my focus on Australia in this chapter, but it's where I live and know best. In the next chapter we will take a look at tall wooden buildings across the planet.

CHAPTER 14

# An international phenomenon

Tall buildings with the resilience of trees

Around the world many building codes do not now discriminate based on materials alone, preferring a judgment focused more on function. Traditionally multi-storey projects in Europe and the United Kingdom have not used wood due to engineering concerns and its inability to meet acoustic and fire safety ratings.

Mansion, Sinaia, Romania, substantial timber construction, with carbon storage and other environmental credentials.

However, more recently the United Kingdom has seen an increase in multi-storey timber buildings, especially for residential and education purposes. In part this has been due to carbon storage and other environmental credentials imposed by various levels of government.[147]

•

Changes to the building code in the Canadian province of British Columbia were instigated in 2009. These changes raised the maximum height for wood-framed residential buildings from four to six stories. Also the Canadian *Wood First Act*, was enacted that required wherever possible government-funded building projects to be constructed from wood.

Code changes began a rapid increase in the number of wood-constructed buildings, with over 250 either completed or begun during the decade following the code change. The success of these efforts has motivated other Canadian provinces to enact similar reforms to their building codes.

By 2010, two Scandinavian countries also changed their building codes to allow taller residential structures built with timber framing, or out of engineered wood products, such as CLT. The Swedish code allows for apartments of up to eight stories to be built from wood. Finland followed suit in changing their codes to more closely align with their neighbour, allowing for timber-frame structures also up to the same eight story mark as long as suitable fire-resistance measures were adopted.

These examples from different countries show a distinct trend for tall buildings from concrete and steel to structures of conventional or engineered wood product timber-framed

---

147 *Code changes and multi-storey timber – a brief international experience* 5 February 2016. See: www.woodsolutions.com.au

An international phenomenon **147**

construction. Changes to the building codes often do not come alone, but are accompanied by a corresponding encouragement from governments towards builders and designers to use more wood in construction.[148]

There are the same regulatory requirements on wooden buildings as there are for other building systems. It is notable that unlike other materials, wood behaves predictably in fire, forming a charred surface which provides protection for the inner structure, so that wood elements can remain intact and fully load-bearing during a fire. Modern timber buildings readily comply with sound insulation standards through using a layered structure of different materials. Even more demanding standards can be met using a number of different design solutions, setting no limits to carbon efficient construction in wood.

Residential building Hummelkaserne, Graz, Austria: Four similar six storey high residential buildings with a larch façade: More advanced wood building systems are continually being developed and refined.

More advanced wood building systems are continually being developed and refined, so around the world we see building codes shifting to avoid discrimination based on material, and instead

---

148 *Code changes and multi-storey timber – a brief international experience* 5 February 2016. See: www.woodsolutions.com.au

judged solely on performance. So for example Australia has now joined others countries allowing their economies and environments to benefit from the advantages of multi-storey timber construction.

●

Interestingly there now appears to be a tall wooden buildings race with numerous proposals and contenders on drawing boards.

The tallest timber tower yet has been proposed for central London, overlooking St Paul's Cathedral. This sleek, 80-storey tower will soar 300 metres above central London's famous Barbican area, and will be a 95,000 square metre mixed-use building, including 1000 residential units. The Oakwood Tower or Toothpick, as it is being called, will be made almost entirely from wood.[149]

Dr Michael Ramage, Director of Cambridge's Centre for Natural Material Innovation said that if London was going to survive it needed to increase its density. "We believe people have a greater affinity for taller buildings in natural materials rather than steel and concrete towers. The fundamental premise is that timber and other natural materials are vastly underused and we don't give them nearly enough credit."[150]

He insists that the timing is right for projects like the Toothpick, that he designed with engineers from London's PLP Architecture. "While the latest steel-and-concrete skyscrapers may look sleek and cutting-edge, most come with old-world energy bills."

"Timber has a very important place in the future construction of medium and large-scale buildings. It can be cost-effective, faster to build, and result in more attractive high-rises," said Ramage.

He believes the most likely roadblock ahead for tall wooden

---
149 *London's 80-storey wooden skyscraper could solve a major architectural problem* Adam Payne, 12 May 2016 Business Insider.
150 *Architects want to build a wooden skyscraper in the centre of London* 11 April 2016 Duncan Geere TechRadar Newsletter www.techradar.com/news

An international phenomenon  149

Oakwood Tower or 'Toothpick': An 80-storey tower that will soar 300 metres above London's famous Barbican area and be made almost entirely from wood. It will be a 95,000 square metre mixed-use building, including 1000 residential units.

towers is not engineering, but public attitude. He admits the idea of living in a timber skyscraper takes some getting used to. "It's difficult to get people out of the mindset that steel or concrete are intrinsically more secure. But isn't it great that the building material showing the most promise in the world today is the one that humans have had an affinity with since we started looking for shelter beyond the caves?" said Ramage.

Kevin Flanagan, Partner at PLP Architecture, said that we now live predominantly in cities, and so the proposals have been designed to improve our well-being in an urban context. "Timber buildings have the potential architecturally to create a more pleasing, relaxed, sociable and creative urban experience. Our firm is currently

Brock Commons Student Residence, University of British Columbia. An 18-storey $50 million-plus hybrid building under construction in Vancouver, Canada is designed as student accommodation. When completed the building will be the tallest mass wood hybrid building in the world.

designing many of London's tall buildings, and the use of timber could transform the way we build in this city."[151]

Swedish studio Tham & Videgrd Arkitekter has drawn up plans for a row of wooden multi-storey apartment blocks in Stockholm and French architect Jean Paul Viguier revealed a competition-winning proposal for a trio of timber towers in Bordeaux that will rise to 57 metres. The French government is also backing a cross-industry grouping called Adivbois to build 'exemplar' multi-storey wood residential buildings across France.

A new hybrid building under construction in Vancouver, Canada is designed to house University of British Columbia students. The $50 million-plus project will rise to a height of 53 metres. Construction on the 18-storey Brock Commons Student Residence at the University of British Columbia began in November 2015 and

---

[151] PLP Architecture See: www.plparchitecture.com/plp%2c-cambridge-university-and-smith-and-wallwork-present-timber-skyscraper-research-to-london%e2%80%99s-mayo.html

is expected to be completed in the summer of 2017. The student housing structure will be the tallest mass wood hybrid building in the world. The hybrid construction includes a one-storey concrete podium, two freestanding concrete cores for lateral stability, and 17 storeys of mass timber topped with a prefabricated steel beam and metal deck roof. Steel connectors will be fitted to the building's glulam columns to provide direct load transfer between the columns and a grid of CLT panels, allowing the building to meet new seismic design requirements for the National Building Code of Canada.

The successful completion of Brock Commons student residence is expected to increase the acceptability of tall timber constructions as well as bring about revisions in British Columbia's building codes for tall wood structures.

---•---

Greater use of timber in construction – together with more trees, parks, green roofs and vertical gardens – will also reduce the urban heat island effect (the phenomenon of higher temperatures experienced in cities due to concrete and tar soaking up thermal energy and radiating it back into the atmosphere) Thanks to better insulation, solar panelling, vegetated roofs, waste-water recycling and low-watt lighting[152], this effect can be substantially off-set.

A report by the UN Food and Agriculture Organisation[153] calls for more wood and wood-based materials to be used in construction instead of non-renewable materials like concrete, brick and steel. The report says wood products and wood energy can replace fossil-intense products in other sectors, creating a virtuous cycle towards low-carbon economies. A key message of the report is:

---

152 *New wood: how it will change our skyline. Timber buildings are reaching towards the skies, thanks to breakthroughs in super-strong wood*, 27 August 2016 Greg Callaghan, Sydney Morning Herald Good Weekend Magazine.

153 *Forestry for a low-carbon future: Integrating forests and wood products in climate change strategies* 2016 Food and Agriculture Organization of the United Nations, FAO Forestry Paper 177

*Increased use of wood offers important mitigation potential when it displaces fossil-fuel intense products. Production of wood-based materials and products results in lower greenhouse gas emissions than production of other materials such as concrete, metal, bricks and plastic. Responsible management of end-of-life wood products, as well as of other biomass residues generated along the wood product value chain, is critical to ensuring a low carbon footprint.*

---

A construction challenge for all skyscrapers, but particularly lighter, timber ones, is swaying. Wind pushes against structures and accelerates upward in what's known as the stack effect. One solution to reduce sway is to add concrete elements in the middle storeys and roof.

Saving an enormous amount of carbon emissions if the bulk of the building is built in wood.

"A timber high-rise of 25 storeys is manageable and feasible," says Andrew Nieland, an architect with Lend Lease, the company behind Melbourne's Forté and Sydney's International House. "Beyond that, you need a hybrid building, one with concrete or steel inclusions. Even so, you'll still save an enormous amount of carbon emissions if the bulk of the building is timber."

That very lightness of timber can be an advantage in seismically active areas if the buildings are well constructed (in Haiti, Nepal and Japan, many timber buildings remained standing amid a sea of grey rubble in the wake of major earthquakes). And timber high-rises may have a role to play in areas of soft soils (Shanghai, for example, has sunk 40 centimetres in the past 50 years, a legacy of its soft soils and global warming, possibly made worse by the weight of its vast mountain of skyscrapers).

"Timber buildings have the resilience of trees; they can creak and move. Concrete buildings have a cracking point; once they start

The construction sector holds a the potential for lowering greenhouse gas emissions as wood-based building materials have several direct and indirect climate related advantages.

to crumble, they have to be rebuilt," observes Dylan Brady, chief architect with Decibel Architecture in Victoria, Australia. Brady has designed an eight-storey CLT building in Punt Road, Melbourne, which he describes as "townhouses in the sky".[154]

•

The construction sector holds a significant potential for lowering greenhouse gas emissions. As new buildings are becoming more energy efficient the energy demand and carbon footprint of construction materials is seen as becoming increasingly important. Wood-based building materials have several direct and indirect climate related advantages. However, strong political actions will be needed in order to utilise this potential for the mitigation of climate change. The bottom line is that we need to protect the climate we all enjoy – and the fragile planet we all share.

---

154 *New wood: how it will change our skyline. Timber buildings are reaching towards the skies, thanks to breakthroughs in super-strong wood*, 27 August 2016 Greg Callaghan, Sydney Morning Herald Good Weekend Magazine.

CHAPTER 15

# More wood wonders

Biology and renewability keys to
continuing benefits

Having already tracked across the benefits – existing and emerging – of wood and wood products for building and construction, this chapter will attempt to round up related benefits. Then in the next chapter we will concentrate a bit more specifically on the use of wood as a source of renewable energy. In this chapter we will focus on some of the benefits using wood products will bring to the climate change challenge. These benefits add to those already discussed, helping to strengthen the already strong case.

In the previous chapter, we tracked through the significant $CO_2$ savings in greenhouse gas emissions, embodied energy savings, and in energy efficiency terms to be made by using wood products in the construction of housing and other buildings with a fair bit of attention to the emerging, exciting tall wooden building opportunity. The good news doesn't stop there. In most cases at the end of their service life, wood products can be recycled, extending their carbon storage credentials. Or alternatively, they can be used as a carbon neutral fuel, substituting for fossil fuel sources.

•

I think we have already made it clear that there are just two ways to reduce $CO_2$ in the atmosphere – either by lowering emissions, or by removing $CO_2$ from the atmosphere and storing it. So either reducing carbon sources or increasing carbon sinks. We have already said that because of its biological origin and renewability, wood

Timber at warehouse. Significant savings in greenhouse gas emissions, embodied energy savings, and in energy efficiency terms to be made by using wood products in the construction of housing and other buildings.

products have the ability to do both. Rather than carbon sinks, as they do not themselves capture $CO_2$ directly from the atmosphere, wood plays an important part in enhancing the effectiveness of forest and tree plantation sinks by extending the period that the $CO_2$ captured by trees via the process of photosynthesis is kept out of the atmosphere, and by encouraging increased tree planting and growth.

As long as the $CO_2$ remains stored in wood, any increase in the volume of 'wood storage' will reduce the amount of $CO_2$ in the atmosphere. So obviously increasing the use of wood from sustainable forest and plantation sources is a simple way of reducing climate change impacts.

In a Harvard Project on Climate Change paper[155] René Castro-Salazar[156] concludes that agriculture, forestry and land-use activities:

---

[155] *Eco-Competitiveness and Eco-Efficiency: Carbon Neutrality in Latin America* 2015, Harvard Project on Climate Change.

[156] René Castro-Salazar is the Assistant Director General, Climate, Biodiversity, Land and Water Department, FAO and previously a Fellow, Mossavar-Rahmani Center for Business and Government, Harvard Kennedy School. Prior to that he was the Minister of Environment, Energy, and Telecommunications, Government of Costa Rica.

More wood wonders  **157**

René Castro-Salazar, Assistant Director-General, Climate, Biodiversity, Land and Water Department. Firmly and forcefully of the view that forests are at the heart of a transition to low-carbon economies.

*... have the greatest potential for carbon sequestration and offer abatement opportunities that are cheaper than can be found in the energy or transport sectors. In Central America, about 80% of agriculture, forestry, and land-use carbon mitigation options cost far less than $20/ton.*

---

When wood cannot be reused, it can be deployed to produce energy through combustion. The energy produced from such combustion is effectively stored energy from the Sun. As the amount of $CO_2$ emitted from the combustion process is no more than the amount previously stored, burning wood is carbon neutral. I think that makes sense.

Using wood also helps to save energy over the life of a house or building. Its cellular structure provides outstanding thermal insulation – fifteen times better than concrete, 400 times better than steel and 1770 times better than aluminium. A two-and-a-half

Landfill wood waste as an energy source will reduce dependence on fossil fuels. The energy produced from such combustion is effectively stored energy from the sun.

centimetre wooden board has better thermal resistance than an 11.4 centimetre brick wall.[157] As a result, wood is becoming an ever more competitive solution to the increasing thermal demands of house and building regulations. So it is clear that the naturally good thermal insulation attributes of wood make it the building material of choice, both in cold and hot climates.

Another positive attribute, unlike other materials, is that wood behaves predictably in fires, forming a charred surface providing protection for the central wood 'core'. Wood products can therefore remain intact and load-bearing during a building fire.

Professor of Fire Safety Engineering at the University of Queensland José Cullen has been testing fireproofing in wooden

---

157 Timber Research and Development Association, UK www.trada.co.uk

Unlike other materials, another positive attribute is that wood behaves predictably in fires, forming a charred surface providing protection for the central wood 'core'.

buildings. In relation to mass timber product technology, he asserts that if a fire begins in the outer layers of the CLT, panels will only char, protecting the core. This means that if a timber high-rise did ignite, the structural integrity of the building should be maintained. He notes that steel buckles under extreme heat: "turning to spaghetti at about 260°C."[158]

Using a layered structure of different materials, modern timber buildings readily comply with sound insulation standards. Even more demanding standards can be met using a number of different design solutions, setting few limits to carbon-efficient construction in wood.

The energy required in the manufacture of building materials is becoming an increasingly important consideration in the climate change debate. As we have discussed, building and construction industries hold significant potential for lowering greenhouse gas emissions.

---

158 Professor José Torero Cullen, School of Civil Engineering University of Queensland. See: http://researchers.uq.edu.au/researcher/2879

Interior, CLT construction: Environmental rating schemes provide an opportunity for wood products to demonstrate their climate change and other environmental credentials.

The energy required to manufacture building products is becoming an increasingly important consideration as more and more architects and home owners consider the environmental impacts of the products and materials they select. Where the building material originates from, how it is used or converted into a building product and its use right through to its reuse, recycling or other disposal are now critical issues.[159]

The energy used to create the products that make up a building is typically around 20 per cent of the total energy expended over the lifetime of the building[160], so it is worth paying attention to the materials specified, as well as to the energy efficiency of the structure.

159 *Building Sustainably with Timber* 2004 Building Research Establishment www.woodforgood.com/bwwpdf/bswt.pdf.
160 *Tackle Climate Change: Use Wood* 2009 CEI-Bois www.cei-bois.org

Sustainability rating systems for buildings, such as the Green Building Council of Australia's Green Star rating system[161] are becoming increasingly important to architects, building developers and consumers looking to reduce their environmental 'footprint' and gain a marketing advantage.

Environmental rating schemes can include 'life-cycle assessments' that compare the whole-of-life performance of building products. Such schemes provide an opportunity for wood products to demonstrate their environmental credentials, particularly in relation to avoiding greenhouse gas emissions, and their contribution to climate change abatement.

•

Thanks to photosynthesis, there is no other building material that requires so little energy in its manufacture as wood. Not only is the production and processing of wood products highly energy-efficient, but engineered wood products can often be used to substitute for steel, aluminium, concrete or plastics that require large amounts of energy to produce.

So every cubic metre of wood used as a substitute for another building material reduces $CO_2$ emissions to the atmosphere by an average of about a tonne of $CO_2$. If this is added to the bit under a tonne of $CO_2$ stored in wood, each cubic metre of wood saves around two tonnes of $CO_2$. Based on these figures, a ten per cent increase in the percentage of wooden houses in Europe would produce sufficient $CO_2$ savings to account for about 25 per cent of the reductions for Europe prescribed by the Kyoto Protocol.[162]

•

---

161 Green Building Council of Australia 2015 www.gbca.org.au
162 *Comparison of wood products and major substitutes with respect to environmental and energy balances* ECE/FAO seminar: Strategies for the Sound Use of Wood, Romania 2003.

Disabled timber beach access, Phillip Island, Victoria, Australia. Every cubic metre of wood used as a substitute for another building material reduces $CO_2$ emissions to the atmosphere by an average of about a tonne of $CO_2$.

There is growing international momentum for governments to enact 'wood first' policies in government-funded building and building product procurement policies. This 'wood first reference' has now been implemented by several local government authorities across Australia. In the state of Queensland, Timber Queensland has launched a wood encouragement campaign. This campaign calls on the state and local governments to consider wood as the preferred material of choice for all public buildings and construction procurement projects. The campaign notes that a wood preference policy has so far been implemented in seven regions of the states of Victoria, New South Wales, South Australia and Western Australia.

Working in conjunction with Planet Ark, Timber Queensland[163] chief executive Mick Stephens said the campaign is exciting. "Particularly given recent changes to the National Construction Code, that allow builders to use timber building solutions up to 25 metres, or eight stores."

"By adopting wood encouragement policies, the Queensland Government and local councils can accelerate the uptake and use of timber in mid to high-rise building projects, such as the recent trend in mid-rise housing construction."

"With the population of Queensland alone expected to nearly double by 2060 a significant amount of new housing and community infrastructure will be required. A wood encouragement policy would reduce the carbon footprint of the state's built environment into the future, including such facilities as schools, libraries, hospitals, retail shops, fire stations, bridges and wharfs'" said Stephens.

He asserted that switching to the greater use of wood products in buildings would generate significant carbon benefits. "If half of all new residential dwellings built in Queensland in any one year were 'timber maximised', this would create a saving of 600,000 tonnes of carbon emissions a year compared to other materials, or six million tonnes over a ten year period."

He said the increased use of timber also has the added benefit of supporting regional industries in Queensland, which directly employ more than 13,000 people across the state, and 21,000 jobs indirectly.

Planet Ark's *Make it Wood* program manager David Rowlinson confirmed that the building sector can play a major art in helping to meet Australia's Paris Agreement commitments. "We know that buildings produce almost a quarter of Australia's greenhouse gas

---

163 Timber Queensland is the peak timber industry body in Queensland and has a key role of supporting and encouraging the development and expansion of the forest and timber industries as a means of securing the long-term viable business activity.

emissions. The increased use of wood, which consumes significantly less energy than traditional materials like concrete and steel, can play a key role in significantly reducing these emissions."

"Also wooden buildings provide many health benefits over other building materials options. Research has revealed physiological and psychological health benefits, including reduced blood pressure, heart rate and stress levels and improved emotional state and level of self-expression," he said.

———————————— ○ ————————————

Contrary to the belief that there is a causal link between using wood and the destruction of forests, increasing the use of wood makes a positive contribution to maintaining and increasing forests and tree plantations, as it creates a market value for forests that is a powerful incentive to perpetuate them, and to establish further plantations. Plus any increase in the global volume of 'wood storage' will reduce the $CO_2$ in the atmosphere, so increasing the use of wood is a simple way of reducing climate change.

The saying that 'a forest that pays is a forest that stays'[164] may be a simplification, but it illustrates a simple truth – a forest's long-term survival depends on its value to the community and economy. Developing a market for wood helps owners and governments see forests in a positive light that recognises their contribution to local and national economies. When community prosperity is associated with the presence of a forest, the principles of sustainable management are respected and the enthusiasm for additional tree plantations increases.

In drawing this chapter to a close we should acknowledge that despite the obvious advantages of wood and its potential to mitigate

---

164 Commonwealth of Australia Minister for the Environment, Hon Greg Hunt, 2014 Asia-Pacific Rainforest Summit, Sydney 2014.

Log harvesting, pine plantation, Australia. A forest's long-term survival depends on its value to the community and economy. Developing a market for wood helps owners and governments see forests in a positive light that recognises their contribution to local and national economies.

climate change, there are still legislative barriers and other obstacles that continue to hamper the extended use of wood and wood-based products in building and construction. Strong political action will be needed in order to utilise the potential that trees and wood products offer for the mitigation of adverse climate change.

CHAPTER 16

# Burning wood

Exploring the contradiction that burning trees fights climate change

We all know that wood burns – even if at times it can be hard to get a fire started! Also getting large pieces of wood to burn can be difficult. Nevertheless, the fact that wood does burn provides a further dimension to the already strong case that trees and wood are important weapons in the climate change challenge. We are going to explore this apparent contradiction in this chapter.

Since humanity's early days on the planet, wood, in the form of fuel and heat for cooking and primitive manufacturing, was the first source of energy. This still remains the situation in many countries. For instance, 80 per cent of the wood harvested in

Since humanity's early days on the planet, wood, in the form of fuel and heat for cooking and primitive manufacturing.

Africa is still used for household cooking and heating, plus some basic industrial applications.

However, the industrial revolution in the developed world saw coal, oil and gas supersede wood as the fuel of choice to produce heat, steam and electricity for 'modern' industrial processes. Although, as we will see, wood is making a comeback, and is set to be central to our renewable energy future.

———————————— • ————————————

To get us started let us look at a bit of the background about 'traditional' current day fuels. Coal had its origins in ancient plant life. Much of the coal that exists today began its formation around 300 million years ago, when the Earth had a rich mantle of forests and swamps. As trees died, peat, comprising about 90 per cent water, was formed. Over time this peat was buried in sediment, compressed and eventually fossilised into coal. This is why coal is referred to as *fossil fuel*.

Today the great bulk of energy and fuel supplies are still produced from fossil fuels. As demand overtakes supply capacity these energy and fuel supplies will continue to increase in price. On a positive note climate change-wise, the World Energy Outlook 2014[165] expects renewable energy to account for nearly half of the global increase in power generation by 2040, and to overtake coal as the leading source of electricity at about that time.

———————————— • ————————————

If we are going to talk about burning wood as a source of energy, let us be clear about the terminology before we go any further. Wood is a subset of biomass. Biomass is organic matter derived from plants through the process of photosynthesis that we have

---
[165] International Energy Agency (IEA), 2013 *Key World Energy Statistics*. Sourced from: www.iea.org/publications

already considered in detail. So biomass is essentially renewable energy for producing power, heat and transport fuels that results in a negligible net contribution of $CO_2$ to the atmosphere. This is because the amount of $CO_2$ emitted from combustion of biomass is no more than the amount previously stored. So burning wood is carbon neutral. Put rather better than I could by Peter Wohlleben in his book *The Hidden Life of Trees*[166]:

> As they photosynthesise, they produce hydrocarbons, which fuel their growth, and over the course of their lives, they store up to 22 tons of carbon dioxide in their trunks, branches and root systems. When they die, the same exact quantity of greenhouse gas is released as fungi and bacteria break down the wood, process the carbon dioxide, and breathe it out again. The assertion is that burning wood is climate neutral is based on this concept. After all, it makes no difference if it's small organisms reducing pieces of wood to their gaseous components or if the home hearth takes on this task, right?

Biomass includes wood from forests, tree plantations and wood processing facilities, plus residues from agricultural crops, organic waste from industry, and food production. Biomass-derived energy can be utilised on-site to help meet manufacturing heat and power needs, whilst at the same time avoiding waste disposal costs.

166 *The Hidden Life of Trees* Peter Wohllben, 2016 Black Inc, Victoria, Australia Page 93.

Plant biomass for energy production is presently available as residues from processing operations, such as sawmills and sugar mills. Biomass includes wood from forests, tree plantations and wood processing facilities, plus residues from agricultural crops, organic waste from industry, and food production. Examples are woodchips, sawdust, wood processing off cuts, cotton ginning trash, nut shells, straw and food waste. Biomass-derived energy can be utilised on-site to help meet facility manufacturing heat and power needs, whilst at the same time avoiding waste disposal issues and costs. Surplus power can be sold to electricity distribution grids, and surplus heat to nearby users.

---

The production of electricity and liquid fuels from wood is gaining increased attention in the global forest and wood products sector. Biomass can be readily adapted to current fossil fuel power production technologies and is very competitive with wind, hydro and solar sources of renewable energy.

It may surprise you to know that the cost of conversion to electricity is essentially the same for biomass and coal. The difference, and it can be significant, in the overall price relates to the cost of collecting and transporting biomass.

So, as we have said, the conversion of biomass to energy – or bioenergy – is renewable and carbon neutral. It is carbon smart compared to the use of fossil fuel-derived energy. $CO_2$ released during the energy conversion process circulates in the atmosphere and is reabsorbed in new biomass growth created through that miraculous process of photosynthesis.

As we have outlined, the making of energy from biomass has a well established track record of cost effectively reducing carbon

The production of electricity and liquid fuels from wood is gaining increased attention in the global forest and wood products sector. Biomass can be readily adapted to current fossil fuel power production technologies.

emissions, improving energy productivity and generating reliable base load[167] energy.

Now for the sad part, even though the sugar industry in Australia has generated electricity and heat from sugarcane waste for more than a hundred years, bioenergy technologies have not been widely adopted in Australia – contributing less than one per cent of Australia's electricity output. This is well below the OECD[168] average of 2.4 per cent.

Presently about 60 per cent of bioenergy power generation capacity in Australia uses bagasse – the fibrous residue of processes

---

167 Base load power sources are generating stations which can consistently generate the electricity needed to satisfy minimum demand. That demand is called the *base load requirement*, and is the minimum level of demand on an electricity distribution grid. Examples of base load plants include coal fired or nuclear fuelled facilities.

168 OECD: The Organisation for Economic Co-operation and Development is an international economic organisation of 34 countries, founded in 1961 to stimulate economic progress and trade.

sugarcane – as a feedstock. About a further quarter uses biogas produced from urban landfill and waste water treatment plants.

———————— • ————————

Disappointingly, bioenergy generation capacity has grown slowly in Australia and most existing bioenergy facilities are small, with plants generally having generating capacity of less than 10 megawatts (MV).[169] Bioenergy production has increased by an average of just four per cent a year over the last decade, compared with average annual growth of residential solar photovoltaic cell[170] capacity of more than 60 per cent. Wind energy has grown rapidly in Australia, with a present installed capacity of approximately two gigawatts (GW)[171], providing about 1.5 per cent of Australia's electricity needs.

The pedestrian take-up of bioenergy reflects poorly on Australia – a consequence of a combination of cheap coal and lack of government affirmative policy support. It is doubly disappointing considering that Australia produces substantial quantities of wood processing, and forest and tree plantation residues.

Large quantities of these residues have traditionally been exported to Asia and Japan as woodchips for pulp and paper manufacturing. However, this market is now much reduced, as pulp and paper manufacturers prefer to use fast-grown eucalyptus and acacia plantation wood chips sourced from Southeast Asia and Asia. Fibre from these plantations is considered to be of superior quality for paper making than fibre derived from Australian mixed species, older natural forest woodchips.

---

169 Megawatts (MV) are used to measure the output of power plants or the amount of electricity required by a town or city. One megawatt (MW) = 1000 kilowatts = 1,000,000 watts. For example, a typical coal-fired power plant is about 600 MW.

170 A solar cell, or photovoltaic cell, is an electrical device that converts the energy of light directly into electricity by the photovoltaic effect, which is a physical and chemical phenomenon. It is a form of photoelectric cell, defined as a device whose electrical characteristics, such as current, voltage, or resistance, vary when exposed to light.

171 Gigawatts (GV) measure the capacity of large power plants or multiple power plants. One gigawatt (GW) is the equivalent to 1,000 megawatts, or a billion watts.

Power generation wind farm on Royd Moor, South Yorkshire, England. Wind energy has grown rapidly in Australia, with present installed capacity of about two gigawatts.

Within natural forests there are substantial quantities of biomass available that could be utilised as bioenergy feedstock. This material is in the form of log harvesting residues that are presently simply left on the ground or sometimes burnt to reduce fuel levels to decrease the risk of wildfires or to encourage seedling regeneration. Other potential sources of woody biomass include tree plantation thinnings, sawmill residues, urban wood waste and introduced woody weeds, such as camphor laurel.[172]

Where markets for bioenergy presently exist or arise in the future, the income received from the sale of this biomass is, or will be, additional to conventional log sales. This additional income will

---

[172] Camphor laurel (Cinnamomum camphora) is an evergreen tree that grows up to 20 metres. It has a large, spreading canopy and a short, stout trunk up to one and a half metres in diameter. Camphor laurel is easily identified by the pungent camphor odour arising from crushed leaves or exposed wood.
   Camphor laurel is categorised as a weed species in Australia. It has the ability to adapt to disturbed environments. It has prolific seed production and rapid growth rate as well as a lack of serious predators or diseases. On the north and mid north coasts of New South Wales, camphor laurel invades large areas of land and inhibits potential land other uses. The contraction of dairying and banana farming since the 1960s has resulted in large areas becoming infested with camphor laurel.

Forest and wood processing residues stored at biofuel power plant.

improve tree utilisation and enhance the efficiency of forest and plantation-based businesses.

Further, utilising wood processing and manufacturing residues as a source of bioenergy provides a carbon neutral substitute for fossil fuels. As such bioenergy only returns to the atmosphere the $CO_2$ that has been previously absorbed by the growing trees. So wood combustion does not contribute to the global warning.

•

In Europe the use of biomass for energy production has increased significantly over the last ten years or so, and ambitious plans exist for further expansion to achieve bold renewable energy targets. Across Europe and Scandinavia there is widespread acceptance that forest biomass from sustainable forest management activities can significantly reduce carbon emissions, create regional economic growth and produce other benefits, such as decreased fire risk. Sweden, Finland and the Baltic countries of Latvia, Lithuania and

**174** By the light of the Sun

Wood chips for bioenergy use. There are substantial quantities of biomass available that could be utilised as bioenergy feedstock. This material is in the form of log harvesting residues that are presently simply left on the ground or sometimes burnt to reduce forest fuel levels. Other potential sources of woody biomass include tree plantation thinnings, sawmill residues, urban wood waste and introduced woody weeds.

Biomass storage. Across Europe and Scandinavia there is widespread acceptance that forest biomass from sustainable forest management activities can significantly reduce carbon emissions, create regional economic growth and produce other benefits, such as decreased fire risk.

Estonia, all with substantial forest estates, produce about a third of their energy consumption from mostly forest-sourced biomass.

Despite the intensified use of forest biomass for energy, the standing volume of European forests increased 12 per cent in the last ten years.[173] This demonstrates that the use of lower quality wood from forests and wood processing residues can make a significant contribution to renewable energy targets.

———————— • ————————

But hold on – make no mistake – the use of forest and tree plantation biomass for energy is not a panacea, although its use can make a worthwhile contribution to renewable energy generation, climate change mitigation and regional economic development.

However, as already indicated, the brutal reality is that in many situations more than fifty per cent of the cost of producing energy from biomass is attributable to the costs of harvesting, transporting, and 'conditioning' the biomass prior to conversion to energy. The high bulk and low density of most biomass feed stocks means that volume, rather than weight, invariably limits transport options and elevated costs.

———————— • ————————

Bioenergy also contributes to Australia's current transport fuel consumption. Ethanol and biodiesel are manufactured from materials like molasses, waste starch, tallow and used cooking oil. In total they provide approximately two per cent of Australia's current transport fuel.

In the longer term, bioenergy could provide 20 per cent or more of Australia's power and transport fuels. Such increased bioenergy use could be supported by the development of sustainable biomass

---

173 European Biomass Association (AEBIOM) 2012 European Bioenergy Outlook, Brussels http://www.aebiom.org

Bioenergy contributes to Australia's current transport fuel consumption. Ethanol and biodiesel are manufactured from materials like molasses, waste starch, tallow and used cooking oil, but they provide just two per cent of Australia's current transport fuel.

resources. Particular opportunities involve existing agricultural residues, future tree plantings, both for biomass, and for other environmental benefits.[174]

———————————— • ————————————

The commercial development of technologies for conversion of biomass into energy products is a critical part of bioenergy development. There is already a range of clean energy technologies in use, or in an advanced stage of development for generating electricity from biomass. These technologies include direct combustion, co-firing with coal, gasification, pyrolysis and fermentation.

Bioenergy plants can range from small domestic heating systems

---

[174] *Bioenergy in Australia Status and Opportunities* 2012, Colin Stucley, Stephen Schuck, Ralph Sims, Jim Bland, Belinda Marino, Michael Borowitzka, Amir Abadi, John Bartle, Richard Giles, Quenten Thomas. Enecon Pty Ltd, PO Box 175 Surrey Hills, Victoria 3127

to multi-megawatt industrial plants requiring several hundred thousand tonnes of biomass feed-stock a year. Presently about 90 per cent of the world's bioenergy plants use combustion processes. The maturity of combustion technology is evidenced by more than 70 GW of installed bioenergy capacity globally for electricity generation alone.[175] Conventional combustion technologies create heat and can therefore generate steam. This steam can then be utilised through a conventional turbo-alternator to produce electricity.

Cogeneration technologies involve the production of electricity, plus heat, and improve the overall thermal efficiency of combustion technologies. Co-firing of biomass with fossil fuels, such as coal,

Bioenergy plant with storage of wood fuel. The development of technologies for conversion of biomass into energy products is a critical part of bioenergy development.

---

175 *World Energy Outlook 2012* Chapter 7 Renewable energy outlook 2012 International Energy Agency. See: / www.worldenergyoutlook.org

offers a relatively low cost bioenergy combustion option, in that existing fossil fuel power stations and related infrastructure can be used without the need to construct new facilities.

Emerging thermo-chemical technologies for biomass conversion to energy are gasification and pyrolysis. Gasification is essentially a process that takes place in a restricted supply of air or oxygen and produces a fuel gas from biomass that is rich in combustible carbon monoxide and hydrogen. This gas has a lower calorific value than natural gas, but after cleaning of tars and particulates can be used as fuel for boilers, engines and combustion turbines.

Pyrolysis technologies involve the thermal degradation of biomass in the absence of air or oxygen. This process removes hydrogen and oxygen so that the remaining char is composed primarily of carbon. This converts the biomass into a combination of solid char[176], gas and liquid, called bio-oil that can be used as a liquid fuel.

---

In this chapter I have attempted to explore the present and future opportunities that wood, in the form of biomass, offers to the production of renewable energy. Further, that by doing so, contributes to climate change abatement. As outlined, at least in Australia, this opportunity is yet to be fully realised, even though the biomass is present and the technology exists, or is rapidly emerging. Sure, cheap fossil fuel in the form of coal and government policy ambivalence are challenges. Also the collecting and transport of biomass remain serious commercial challenges if bioenergy is to achieve its real potential in contributing towards renewable energy targets and tackling climate change.

---

176 Produced by incomplete combustion in the presence of high heat. The resulting residue is called char.

CHAPTER 17

# Action stations

Harnessing the power of photosynthesis-derived products

I want to start this concluding chapter by summarising some of the salient points from preceding chapters. Then I will endeavour to outline actions to address climate change that have their foundations in the miracle of photosynthesis. More specifically, in the context of this book, storing carbon in the wood of tree trunks. So whilst acknowledging, as we have done already, that it is but one of the essential actions to combat climate change, the mission here is how to better harness the power of the products of photosynthesis to offset climate change impacts. Actions that can be undertaken right now, and don't require great innovation or future technology.

Back in Chapter 1, I affirmed that this book was not going to debate whether or not the climate is changing. I asserted that the scientific argument is settled, and that adverse climate change is the overriding environmental issue facing the survival of humanity. The central purpose of this book I said, and say again, is to attempt to outline what we can do to slow, and eventually halt this human-induced environmental juggernaut. What can we do to better understand and use the process of photosynthesis, and the central role trees can play in the fight to arrest catastrophic climate change?

⸺⸺⸺⸺ • ⸺⸺⸺⸺

We have noted the turbulence in public and media discussions in considering climate change because of its political, commercial and industrial impacts, plus seen a proliferation of misleading

Understanding the process of photosynthesis, and the central role trees can play in the fight to arrest catastrophic climate change.

science and scary tactics, as special interest groups argue their case essentially creating opportunities for political mischief, policy inertia and public confusion.

We have also noted the equivocation, notably by developed countries, regarding commitments to affirmative climate change action to reduce $CO_2$ emissions and to embrace renewable energy technologies. This apparent developed country insincerity is evidenced by the language of the United States Trump administration in relation to the country's commitment to climate change action, including the Paris Agreement. On 7 November 2012, before the last election, now President Donald Trump tweeted:

*The concept of global warming was created by and for the Chinese in order to make U.S. manufacturing non-competitive.*

After Trump assumed the presidency, on 31 January 2017, under the headline: *President Trump Prepares to Withdraw from Groundbreaking Climate Change Agreement*, a *Reuters* article reported:

> The United States will switch course on climate change and pull out of a global pact to cut emissions ...

However, on 2 February 2017, President Trump signed an executive order declaring that climate change is an immediate threat, both to the economy and to national security. The executive order directed government agencies to develop action plans to reduce their carbon footprints. It also lifted the 'gag order' that had been placed on government scientists and directed Congress to pass legislation aimed at reducing carbon emissions in the United States of America by 20 per cent over the next five years. In signing the executive order President Trump said:

> I've spoken with some of the top scientists here in Washington, and let me tell you, I trust these guys. They showed me without a doubt that climate change is real and we need to do something about it now.

President Donald Trump. Hard to be confident about the policy of the United States in relation to climate change action.

However, in a not unexpected move President Trump withdrew the United States of America from the Paris Agreement on 1 June 2017 saying he was taking his decision to protect American jobs. The United States joined Syria and Nicaragua as the only countries to be outside the Paris Agreement.

President Trump claimed that the Paris Agreement allowed countries, such as China and India to carry on polluting while the United States economy was harmed. He said:

*We're getting out ... it is less about the climate and more about other countries obtaining a financial advantage over the US". We don't want other countries laughing at us anymore, and they won't.*

The President claimed the United States could re-enter "an entirely new transaction", but indicated that was hardly a priority, adding: "If we can, great. If we can't, that's fine."

There was an immediate international consensus that the unilateral decision of the United States marked a sad day for the global community. There was a wide opinion that the United States had disgraced itself before the rest of the world and that the decision would hurt the United States and the planet.

World leaders lined up to criticise the President Trump's move. European leaders released a joint statement condemning his decision, and rejecting his assertion that the climate change agreement could be redrafted. German chancellor Angela Merkel, French president Emmanuel Macron and Italian prime minister Paolo Gentiloni said:

*We deem the momentum generated in Paris in December 2015 irreversible and we firmly believe that the Paris Agreement cannot be renegotiated, since it is a vital instrument for our planet, societies and economies.*

In a televised address Emmanuel Macron said:

*I tell you firmly tonight, we will not renegotiate a less ambitious accord. There is no way. [The President has] committed an error for the interests of his country, his people and a mistake for the future of our planet ... Don't be mistaken on climate; there is no plan B because there is no planet B.*

Angela Merkel called President Trump's decision to withdraw from the Paris Agreement "extremely regrettable" while vowing to continue fighting climate change.

*The decision of the US President to withdraw from the Paris climate agreement is extremely regrettable, and I'm expressing myself in very restrained terms. To everyone for whom the future of our planet is important, I say let's continue going down this path so we're successful for our Mother Earth.*

Chancellor Merkel emphasised Germany's continued commitment to the Paris Agreement, which she called a "cornerstone" of efforts to protect "creation".

She said there was no turning back from the path that began with the 1997 Kyoto protocol and – until 1 June – had the consent of almost every country in the world:

French president Emmanuel Macron: "Don't be mistaken on climate; there is no plan B because there is no planet B."

German chancellor Angela Merkel: "I say let's continue going down this path so we're successful for our Mother Earth."

*We will combine our forces more resolutely than ever ... to address and tackle big challenges for humanity, such as climate change.*

China's premier Li Keqiang said that fighting climate change was a "global consensus" and an "international responsibility". The British government issued a statement saying that prime minister Theresa May had told Trump of her "disappointment" at his decision and stressed that Britain remained committed to the agreement.

Former United States Vice President Al Gore said:

*Removing the United States from the Paris Agreement is a reckless and indefensible action. But make no mistake: if President Trump won't lead, the American people will ... We are in the middle of a clean energy revolution that no single person or group can stop.*

Scientists said the Earth is likely to reach more dangerous levels of warming sooner as a result of the President Trump's decision because, as the world's second-largest emitter of $CO_2$, the United State's made a significant contribution to rising temperatures.

Former United States Vice President Al Gore: "Removing the United States from the Paris Agreement is a reckless and indefensible action."

Environmentalists described President Trump's decision as a "hugely disappointing" mistake. Chief executive of World Wildlife Fund Tanya Steele said the decision makes it harder for the world to reach a safer and more prosperous future. She said:

*It is hugely disappointing that President Trump is making the mistake in rowing back on the Paris agreement," she said, "Climate change is a very real global issue that affects the successful future of our planet.*

Many business leaders said President Trump has handed the advantage in the key field of renewable energy to rivals in China and to Europe. Australian business leaders joined a global chorus of chief executives criticising the United States' withdrawal from the Paris climate accord, saying it created painful uncertainty and worked against the interests of investors.

The Australian Industry Group, which represents more than 60,000 businesses that employ over a million people, including heavy emitting industries like manufacturing and transport, said President Trump's decision "creates damaging uncertainty for businesses worldwide." Australian Council of Superannuation Investors Louise Davidson said:

> It is disheartening to see a decision like this, by a wealthy industrialised nation, which flies in the face of scientific knowledge and investor concerns.

No doubt the decision of the United States of America is extremely disappointing when there is no doubt that countries, especially developed countries with high $CO_2$ emission levels, like the United States, have a moral obligation to adhere to strong $CO_2$ emission reduction policy positions, carbon-neutral technologies and renewable energy production.

———————————— • ————————————

We all know that talk is cheap and it is action that counts. Implementation of policy is the real measure of a government's commitment. So programs that encourage and support – financially and otherwise – climate change mitigation are the real measure of a government's resolve to tackling climate change. Such policy should include planting more trees, using wood-based products to substitute for less climate-friendly ones, pushing hard to turn off the fossil fuel tap, and much more renewable energy effort, including photosynthesis-derived bioenergy.

Certainly the most dangerous impacts of climate change may still be avoided if we can move towards carbon-based renewable sources. Such renewables will need to provide a substantial proportion of our future energy consumption. This notwithstanding political equivocation and confusion intended to orchestrate delays in advancing progress continue to be of concern.

Reaffirming earlier comments, we have identified valuable actions for reducing energy consumption, and as a consequence, lowering $CO_2$ emissions, including using more wood-based materials that use less energy in their manufacture and that also store carbon, and by doing so restrict the use of high energy materials, like steel and aluminium.

---

Again recapping, we have indicated that climate change trends are expected to increase the intensity and frequency of extreme weather events. Rising temperatures are likely to mean that tropical diseases, like malaria, yellow fever, dengue fever and encephalitis will spread, because the area where climatic conditions are suitable for the mosquitoes, ticks and other insects that carry diseases, is and will continue to expand.

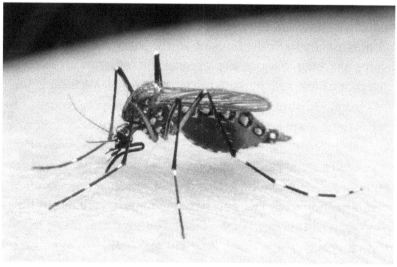

Rising temperatures likely to mean that insect-borne tropical diseases, like malaria, yellow fever, dengue fever and encephalitis will spread.

Yes, most certainly, we know now that reducing greenhouse gas emissions will require substantial changes to how we produce and use energy, but the price of doing nothing will be much higher. The longer we delay the greater will be the damage and human suffering that unconstrained climate change will unquestionably cause.

---•---

It is a 'touchy' subject, but something we simply cannot sweep under the carpet, the excess number of people populating the planet. The size of the human community is now front and centre in the climate change challenge. A finite planet and too many people, and the numbers are continuing to grow. Nearly all our economic, social, and environmental problems are becoming more acute in the face of uncontrolled human population growth.

---•---

We know that photosynthesis stands at the heart of Earth's productivity and that the inverse relationship between respiration and photosynthesis is the basis of all life. Also that each process absorbs the other's waste and excretes the other's raw material converting a lifeless gas into solid, living beings – plants and animals.

As has already been stated there are just two ways to reduce $CO_2$ levels in the atmosphere; either by reducing emissions, or by removing $CO_2$ and storing it – reducing *carbon sources* and increasing *carbon sinks* and that wood has the unique ability to do both.

Of course we know that trees do not just grow on forever. A forest of mature trees 'locks up' large amounts of carbon, but little new growth takes place in such a forest. The important reality worth stressing again in this concluding chapter is that to store ever-increasing amounts of carbon requires trees to be harvested, wood products to be manufactured and tree regeneration or replanting to

Photosynthesis stands at the heart of Earth's productivity. The inverse relationship between respiration and photosynthesis is the basis of all life.

occur. Even then the wood harvested needs to be stored long-term – turned into durable timber products, like houses and furniture. The essential point here is that only by continuing to grow a forest or plantation and removing biomass from time to time in the form of wood can the forest continue to absorb and store atmospheric $CO_2$.

As long as the $CO_2$ remains stored in wood, any increase in the volume of 'wood storage' will reduce the amount of $CO_2$ in the atmosphere. So obviously increasing the use of wood from sustainable forests and plantation sources is a simple way of reducing climate change impacts.

The present and prospective future contribution trees and wood

Mature trees 'lock up' large amounts of carbon, but little new growth occurs. To store ever-increasing amounts of carbon requires trees to be harvested, wood products to be manufactured and tree regeneration or replanting.

products are capable of making to climate change abatement is the central message of this book. The use of forests, particularly well managed tree plantations, to both store carbon and provide an alternative energy source is worthy of greater consideration and application.

———————————— • ————————————

Getting more tree plantations in the ground is not necessarily straight forward. In many countries, from around the 1920s, to perhaps about the 1970s, plantations were established by governments concerned about impending wood supply shortages, or for other reasons, such as employment generation, regional development or utilising land not suitable for agriculture.

Eucalypt plantation, Brazil. The use of forests, particularly well managed tree plantations, to both store carbon and provide an alternative energy source is worthy of greater consideration and application.

Today money spent on planting trees has to have either a compelling environmental purpose, like erosion control or enhancing biodiversity, or more usually a commercial purpose. In the latter case the financial return needs to be sufficiently attractive for investors to want to put their money on the line. Because tree plantations grown to produce wood-based products frequently need 20 or more years to grow to a sufficient size, the combined tyranny of time and compound interest means making an adequate financial return for investors can be challenging. So achieving an adequate return from commercial tree plantation establishment has stalled in some developed economies where a competitive environment exists for investment dollars.

Measures like treating tree plantation proposals more equitably

Treating tree plantation more equitably with other agricultural 'crops', and introducing the ability to create and trade carbon credits can make a worthwhile contribution to the tree plantation investment.

with other agricultural 'crops', and introducing the ability to create and trade carbon credits into the commercial tree plantation financial equation can both make a worthwhile contribution to the tree plantation investment bottom line.

The Australian Carbon Farming Initiative is a scheme that will provide additional economic opportunities for farmers, tree plantation growers and land managers. It will be able to generate carbon credits that can then be sold to other businesses wanting to offset their carbon emissions.

Chair of the Australian Forest Products Association, Greg McCormack, said scientists and foresters have worked closely with the Federal Government to ensure that the carbon storage mathematics in the draft methodology is "bulletproof". He said:

> *Rotational forestry is not just good for jobs in the bush, but also one of the best, and most cost-effective tools for storing carbon and reducing Australia's carbon emissions profile. The additional income will be minor compared to the high costs borne over that first ten-year period. This will ensure that only the most sensible choices are made about where to plant trees – such as close to major processing facilities.*[177]

McCormack added that of course planting trees, whether for landscape repair or commercial purposes provides a range of other benefits, like moderating local climate, biodiversity and aesthetic values.

The opportunity for forests and tree plantations to have a positive role in climate change abatement depends on making progress in arresting tropical forests deforestation and degradation.

---

[177] Australian forestry joins battle against climate change, Australian Forest Products Association media release 2 December 2016. See: http://ausfpa.com.au/media-releases/australian-forestry-joins-battle-against-climate-change

In pursuit of arresting climate change it is obvious that we also need to do all we can to keep existing trees standing, and of particular importance, keep tropical forests intact. To a large degree the opportunity for forests and tree plantations to have a positive role in climate change abatement depends on making progress in arresting deforestation and forest degradation.

To reinforce another fundamental requirement yet again, the development of effective carbon credit trading mechanisms includes the need to be able to quantify the amount of carbon stored in a given area of forest or tree plantation. And to have a legally robust certification system to facilitate the issue of a legal 'property right' that permits carbon credits to be traded and audited, therefore providing adequate assurance for purchasers and government regulators.

---

As has already been mentioned, the Paris Agreement, and the fact that it recognises the key role forests can play in mitigating climate change. We also noted that more than 70 per cent of the countries that have submitted INDCs include forests as an important component of their contribution to climate change action. This includes supporting sustainable management of forests, plus the importance of actions and incentives to safeguard non carbon benefits associated with forest management and protection.

---

In Chapters 13 and 14 mention was made of the exciting times ahead for tall wooden buildings in today's cities, commenting that in many cities population growth, space constraints and infrastructure pressures all contribute to the need for innovative building solutions. In addition, the need for strong environmental credentials in modern

The development of a new generation of strong engineered timber products is taking advantage of timber's dexterity as a building material and its environmental credentials.

cities, is a key driver behind present building activity. An aspect of this reality, that in part takes advantage of timber's dexterity as a building material and its environmental credentials, is the development of a new generation of strong engineered timber products.

Tall wooden buildings weighing up to 50 per cent less than traditional concrete buildings reduce construction costs, particularly where a building site needs extensive outlay on foundations. In dense cities, where conditions are restrained and space is limited, timber construction is becoming increasingly attractive.

There seems little doubt that along with clever design and changes to building codes, engineered timber products and innovative systems will see wooden buildings continuing to move skywards. The

The emerging, exciting tall wooden building phenomenon has the potential for significant $CO_2$ savings in greenhouse gas emissions, embodied energy savings, and energy efficiency.

carbon storage and other environmental attributers of wood will add to this momentum, as will the significant $CO_2$ savings benefits from the use of timber in construction, both in terms of embodied energy and in-use energy efficiency. The construction sector, including the emerging, exciting tall wooden building phenomenon has the real potential for significant $CO_2$ savings in greenhouse gas emissions, embodied energy savings, and in energy efficiency.

---

As already has been noted, the production of electricity and liquid fuels from wood is gaining increased attention, and biomass can be readily adapted to current fossil fuel energy production technologies.

Also importantly, because the amount of $CO_2$ emitted from the combustion of biomass is no more than the amount previously stored, burning wood is carbon neutral and does not contribute to global warming.

So in an increasingly carbon constrained world, forests and wood products are important not only as carbon sinks and to generate carbon offsets, but as substitutes for more carbon intensive materials and fossil fuels.

•

In retrospect I should have said more about the relationship between big business and climate change. It has been, and continues to be mixed, with some business leaders and advocates continuing to be climate change deniers. However, the reality is now rapidly changing. Writing under the subheading of: *The growing spectre of a carbon bubble has prompted changes in the attitudes of big business*, Clancy Yeates suggested that investors controlling trillions of dollars, and powerful regulators have turned their sights on the potentially big financial risks created by a warming climate.[178]

He refers to a report by a group of "corporate giants" and regulators who call for sweeping changes to how much big companies should tell investors about their climate change exposure. He quotes Ross Berry, head of research at the $59 billion find First State Super:

*I think the penny is dropping for everyone that climate change represents one of the most important financial risks for investors. It's not a 50-year thing, it's a here and now thing.*

On a further business-positive note, we have already concluded that moving to a 'low-carbon' economy offers tremendous opportunities for innovation and economic growth. Companies quick to develop

---
[178] Sniffing the breeze to weather rising risks of a changing climate. *Sydney Morning Herald* 25-26 February 2017.

low carbon technologies will have a competitive advantage as global demand for such technologies grows.

I hope I have impressed upon you the urgency of affirmative effort to address climate change. In addition, at a personal level, I strongly suggest reducing expectations and adopting more modest consumption expectations. This might be solar panels, fuel-efficient vehicles, and public transport. Also, please participate in actively encouraging governments to hold the course in terms of policy and its implementation.

As I hope I have been able to demonstrate, there are ways that the light of the Sun, through the process of photosynthesis, can be harnessed to tackle climate change and contribute towards the prospect of an inhabitable planet for our grandkids. If you *really* want to feel good, plant some trees; build or live in a wooden house, or now a wooden apartment – certainly things that you can and should do.

# Acknowledgements

To a notable degree in the early chapters of this book I have lent on the writings of Tim Flannery and David Suzuki. Without reservation I acknowledge their huge contribution to raising environmental awareness and to the climate change debate.

From my perspective, in his books, Tim Flannery demonstrates a bit of a lack of appetite for more trees and increased use of wood as important components of the climate change solution. This view, be it right or wrong, formed part of the motivation for writing this book. I was however delighted to read Tim's foreword in Peter Wohlleben's widely acclaimed book: *The Hidden Life of Trees*.[179] So welcome aboard Tim!

I also feel compelled to acknowledge the outstanding efforts of Forest and Wood Products Australia that provides much of the underpinning of Chapters 13 and 14. Their work in leading changes to the National Construction Code has enabled wood-based buildings in Australia to climb up to 25 metres – really enabling Australia to catch up with the tall wooden building phenomenon sweeping across the Northern Hemisphere.

I really appreciate René Castro Salazar's support in writing the foreword to this book. René is an extremely well credentialed, enthusiastic advocate for tackling climate change and the central role trees, forests and wood products are able to play in this regard. He proved this during his tenure as a senior minister in the Costa Rican Government, as a Fellow at Harvard University, and now in

---

179 *The Hidden Life of Trees*, Peter Wohlleben, 2015, Black Inc, Australia.

his leading role with the FAO of the United Nations. FOA is well served by René's passion, intellectual capacity and leadership. So thanks René!

I should also thank others who have helped to get this book into your hands. Notably Jan Hume provided great assistance with fine tuning the manuscript so it passed muster with the publisher. Russell Jeffery of Emigraph Creative in Sydney provided welcomed help in preparing all the illustrations and layout of the book.

Critically Anthony Cappello and the team at Connor Court Publishing kept the faith and were responsible in having this book designed, published and marketed in both printed and electronic formats.

My grateful appreciation to you all.

# Picture Credits

Every endeavour has been made to identify and attribute the source of pictures. The source of some pictures obtained from the internet has in some cases not been able to be established and acknowledged.

**Anon**
Pages 21, 54, 119, 149, 150, 152, 181

**Australand**
Page 132

**AWISA**
Page 133

**Christoph Kulterer, proHolz, Austria**
Pages 128, 130, 135, 147, 153, 160, 191, 196

**Damien Pleming**
Page 25

**FAO**
Pages viii, 157

**Forestry Tasmania**
Pages 83, 93

**Hardi Baktiantoro**
Page xiv

**John Halkett**
Page 79

**Komatsu Forest Australia**
Page 165

**Lend Lease**
Pages 138, 141

**Maxeine McKeon**
Page xi

**NSW Forestry Corporation**
Page 111

**Shutterstock**
Pages xv, 1, 2, 3, 5, 7, 10, 11, 12,
16, 18, 19, 24, 29, 31, 34, 38, 40,
41, 42, 43, 44, 47, 49, 60, 61, 64,
65, 67, 69, 71, 73, 76, 78, 80, 81,
84, 85, 89, 90, 92, 95, 96, 101,
104, 105, 108, 110, 116, 117,
123, 125, 126, 129, 143, 145,
156, 159, 166, 168, 170, 172,
173, 174 (both), 176, 179, 180,
183, 184, 185, 187, 189, 190,
192, 193, 195

**Stora Enso Australia**
Pages 137, 140

**Timber 2020 Great Southern Private Forestry Development Committee**
Page xiii

**Timber Development Association**
Pages 14, 114, 158, 162

**Wild Light**
Page 63

# Index

## A
Arctic ice melt, 7, 26
    Ocean, 23
    sea ice, 8, 41, 50
atmospheric pollution, xii, 2, 12, 17, 101
Australian Carbon Farming Initiative, 192
Australian Climate Council, 37-39, 42
Australian Industry Group, 185
Australian Sustainable Built Environment Council, 142

## B
bagasse, 170
Barangaroo (International House Sydney), 140, 141
Berry, Ross, 197
bioenergy, 169-173, 175-178, 186
biomass, x, 12, 60, 76, 84-86, 93, 152, 167-170, 172-178, 189,196, 197
Brady, Dylan, 153, 154
Brock Commons Student Residence, 150, 151
Brundtland Commission, 102
building codes, 134, 145-147, 151, 195

## C
cambium 97, 98
carbohydrates, 60

carbon
    credits, 89, 91, 108, 192, 194
    cycle, 17, 57, 93
    dioxide ($CO_2$), x, xii, xiv, xvi, 2, 4, 5, 7, 12-14, 17, 20, 23-26, 39, 46-49, 54-64, 68-74, 76, 77, 87, 88, 91, 96, 97, 100, 101, 103, 125, 128, 130, 139, 142, 155, 156, 157, 161, 164, 168, 169, 173, 180, 184, 186-189, 196, 197
    neutral future, 14
    offsets, vii, 89-91, 94, 197
    prices, 109, 110
    sequestration, viii, 85, 86, 157
    sinks, 57, 60, 62, 63, 73, 94, 111, 155, 156, 188, 197
    Storage, 23, 77, 85, 92, 93, 109, 110, 120-122, 134, 145, 146, 155, 193, 196,
Carson, Rachel, 101, 102
Castro Salazar, Dr René, vii, viii, 156, 157, 199
Chandler, David, 124
chloroplasts, 69
Clean Development Mechanism, 104
clean energy technologies, 176
climate-friendly technologies, 22
coal, 1, 12, 25, 39, 40, 48, 60, 86, 88, 169-171, 176-178
coffee, 8-10
cogeneration, 177
Connor, John, 9

Coral reefs, 40, 41
Costa Rica, vii, viii, 156, 199
Cox, Professor Brian, 6, 7
cross laminated timber (CLT), 133, 135, 137, 138, 140, 141, 145, 146, 151, 154, 159, 160
Cullen, José, 158, 159

**D**
Dandenong Mental Health Facility, 116
Davidson, Louise, 186
DDT, 101, 102
deforestation, viii, ix, xiii, 91, 92, 103, 108, 109, 111, 193, 194
de Rijke, Alex, 124
Doomsday Book, 16

**E**
embodied energy, 125, 155, 156, 196
emissions trading schemes, 88, 89, 91
engineered timber products, 124, 134, 137, 138, 195, 125
Evans, Jonathan, 118

**F**
Farrelly, Elizabeth, 6
Fitzpatrick, James, 135
Flanagan, Kevin, 149
Flannery, Tim, 24-27, 48, 52, 54, 56, 57, 61, 62, 68, 70, 72, 86, 88, 199
Food and Agricultural Organization of the United Nations (FAO), vii, 86
Forest and Wood Products Australia Limited (FWPA), 120, 121, 199
forest categories, 78
degradation, ix, 91, 92, 103, 108, 109, 194

destruction, ix, 55, 88, 100
forester, xi, 192
forests
  boreal, 80, 81
  European, 175
  temperate, 80-84
*Forté* building, 136, 153
fossil fuel, ix, x, 1, 4, 11-13, 23, 25, 39, 48, 55-57, 70, 73, 76, 86, 93, 94, 99, 101, 158, 167, 169, 170, 173, 177, 178, 186, 196, 197
  burning, 1, 12, 23, 25, 55, 57, 88
  emissions, 39

**G**
Gentiloni, Italian prime minister Paolo, 182
Gilbert, Jarrod, 6
Gore, former United States of America Vice President Al, 184, 185
global temperatures, 2, 12, 15, 20, 51, 106
  warming, 18, 25, 26, 42, 72, 105-107, 153, 180, 197
glulam, 135, 140, 141, 151
Green Climate Fund, 112
Green, Michael, 130
Green Star-rated buildings, 144, 161
greenhouse effect, 12, 46, 48, 52, 87, 95
  emissions, vii, 4, 21, 22, 26, 55, 86, 94, 103, 125, 152, 154, 155, 159, 161, 188, 196
greenhouse gases, vi, vii, 4, 12, 21, 27, 46-49, 53, 74, 86, 92, 93, 103, 107, 163, 168

## H

Hageneder, Fred, 67, 68
*Haut* skyscraper, 127, 128
heartwood, 97, 98
Hewson, Nick, 130, 131, 137, 139
Horyuji Buddhist temple, 129

## I

life-cycle assessments, 161
Industrial Revolution, 1, 4, 26, 48, 55, 56, 167
Intended Nationally Determined Contributions (INDCs), 107, 108, 110, 111, 194
Intergovernmental Panel on Climate Change (IPCC), 23, 26, 62, 92, 109
Inuit people, 51, 106

## K

Keenan, Trevor, 72-74
Keqiang, Chinese premier Li, 184
Kyoto Protocol, 103-109, 161, 183

## L

Lawrence Berkeley National Laboratory, California 72
Lend Lease, 136, 138, 140, 153
log harvesting, xi, 77, 165, 172, 174
low-carbon economies, viii, ix, x, 151, 157

## M

McCormack, Greg, 192, 193
McCloskey, Kelly, 133
Macron, French president Emmanuel, 182, 183
Maddison, Peter, 138, 139
*Make it Wood* program, 163
Madew, Romilly, 144
Maher, Professor Ken, 143
Manne, Robert, 17
Marine Fish Farmers Association (Malaysia), 43
mass timber product technology, 159
mass timber, 133, 134, 138, 151
May, United Kingdom Prime Minister Theresa, 184
Merkel, German chancellor Angela, 182-184
Mohamed, Mohamed Razalin, 43
Mooney, Chris, 74
Moore, Sue, 8
Morrison, Ken, 144
multi-storey timber buildings, 133, 134, 146

## N

National Construction Code
    Australian, 132, 134-136, 163, 199
    Canada, 151
National Snow and Data Center, University of Colorado, 50
natural forest systems, xii
Nieland, Andrew, 153
North pole, 23

## O

Oakwood Tower (Toothpick), 148, 149
OECD, 170
organic compounds, 58, 59

## P

Paris Agreement on climate change, x, xvi, 107, 110, 163, 182,-185
photosynthesis, x, xiii, xv, xvi, 4, 14, 48, 49, 57, 59-61, 63, 64, 68-74, 95, 96, 98, 100, 156, 161, 167-169, 179, 180, 186, 188, 189, 198
photovoltaic cells 171
Planet Ark, 113-118, 120, 163
polar bears, 2, 8, 50, 51
prefabricated timber construction, 139
pyrolysis, 176, 178

## R

Ramage, Dr Michael, 148, 149
Reduced Emissions through Deforestation and Degradation (REDD), 89
renewable energy, x, 22, 39, 93, 129, 155, 167-169, 173, 175, 177, 178, 180, 185, 186
rising sea levels, 7, 17, 18, 20, 124
Rowlinson, David, 163

## S

sapwood, 97, 98
Science Summit on World Population, 35
Silent Spring, 101, 102
Sinclair, Ric, 132-134
skyscrapers, 130, 148, 152, 153
Snow, Deborah, 15
sound insulation, 147, 159
South East Asia, ix
Steele, Tanya, 185
Stephens, Mick, 163
Stern, Lord Nicolas, 21, 22, 32, 39

Suhaidul, Maimunah, 45
sustainable forest management, x, xiii, 94, 111, 173, 174
sustainability rating systems, 161
Suzuki, David, 3, 4, 28, 35, 36, 46, 52, 53, 70, 71, 199
Syrian refugees, 19

## T

tall wooden buildings, 124, 129, 144, 148, 194, 195
Taylor, Chris, 125
Tham & Videgrd Arkitekter, 150
The Star, newspaper, 42
thermal insulation, 157, 158
thermo-chemical technologies, 178
Timber Queensland, 162, 163
trees
    African umbrella thorn, 66
    bristlecone pine, 66, 67
    eucalypts, 82, 83, 93, 95, 171, 191
    European oaks, 66
    huon pine, 66
    kauri, 66, 82
    messmate, 66
    monkey puzzle, 82
    mountain ash, 66
    redwoods, 76, 81, 83
tree plantations, xii, 75, 76, 91, 92, 164, 168, 169, 190, 191, 193, 194
TREET tower, 127
tropical forest(s), ix, xiv, 62, 63, 74, 80, 83, 84, 86-89, 90, 100, 103, 193, 194
    forest destruction, ix, 74, 86-90, 100, 103, 193
Trump, President Donald, 180-186

## U

Union of Concerned Scientists, 10, 88
United Nations Conference on the Human Environment, 102
United Nations Framework Convention on Climate Change, 103, 109
United Nations Food and Agricultural Organization, 86
United Nation's population data, 34
Urban Growth NSW, 131

## V

Viguier, Jean Paul, 150

## W

Wade, Matt, 15
Waugh, Andrew, 136
whale population, 7

Wohlleben, Peter, 168, 199
Wood First Act, Canadian, 146
wood first policies, 162
wood hybrid buildings, 150, 151
wood products, ix, x, xii, xv, 4, 5, 13, 14, 74, 75, 84, 85, 93, 94, 97, 107, 115, 118-123, 124, 135, 146, 151, 152, 155, 156, 158, 160, 161, 163, 165, 188-190, 197, 199
Woods, Tim, 133, 134
woodchips, 169, 171, 174
woodscraper, 130
WoodSolutions, 122
World Energy Outlook 201, 167
world population, 28, 30, 35
future population growth, 30, 33, 34

## Y

Yeates, Clancy, 197

Lightning Source UK Ltd.
Milton Keynes UK
UKHW020939130319
339055UK00011B/465/P